NANOTECHNOLOGY FOR MICROELECTRONICS AND OPTOELECTRONICS

Elsevier Internet Homepage

http://www.elsevier.com

Consult the Elsevier homepage for full catalogue information on all books, journals and electronic products and services.

Elsevier Titles of Related Interest

Nanotechnology and Nano-interface controlled electronic devices
M. Iwamoto
0-444-51091-5; 2003; Hardback; 528 Pages

Nanostructured Materials and Nanotechnology
Harwi Nalwa
0-12-513920-9; 2002; Hardback; 834 Pages

Nano And Giga Challenges In Microelectronics
J. Greer, A. Korkin, J. Lababowski
0-444-51494-5; 2003; Hardback

Related Journals
The following journals related to laser processing can all be found at
http://www.sciencedirect.com

Precision Engineering
Microelectronics Journal
Microelectronics Reliability
Microelectronic Engineering
Organic Electronics
Photonics and Nanostructures – Fundamentals and Applications

To Contact the Publisher
Elsevier welcomes enquiries concerning publishing proposals: books, journal special issues, conference proceedings, etc. All formats and media can be considered. Should you have a publishing proposal you wish to discuss, please contact, without obligation, the commissioning editor responsible for Elsevier's materials science books publishing programme:

Amanda Weaver
Publisher, Materials Science
Elsevier Limited
The Boulevard, Langford Lane   Tel.: +44 1865 84 3634
Kidlington, Oxford   Fax: +44 1865 84 3920
OX5 1GB, UK   E-mail: a.weaver@elsevier.com

General enquiries including placing orders, should be directed to Elsevier's Regional Sales Offices – please access the Elsevier internet homepage for full contact details.

NANOTECHNOLOGY FOR MICROELECTRONICS AND OPTOELECTRONICS

J.M. Martínez-Duart
Director of the Department of Applied Physics,
Universidad Autónoma de Madrid,
Madrid, Spain

R.J. Martín-Palma
Universidad Autónoma de Madrid,
Madrid, Spain

F. Agulló-Rueda
Materials Science Institute of Madrid, CSIC,
Madrid, Spain

ELSEVIER

AMSTERDAM • BOSTON • HEIDELBERG • LONDON • NEW YORK • OXFORD
PARIS • SAN DIEGO • SAN FRANCISCO • SINGAPORE • SYDNEY • TOKYO

ELSEVIER B.V.	ELSEVIER Inc.	ELSEVIER Ltd	ELSEVIER Ltd
Radarweg 29	525 B Street, Suite 1900	The Boulevard, Langford Lane	84 Theobald's Road
P.O. Box 211, 1000	San Diego, CA 92101-4495	Kidlington, Oxford OX5 1GB	London WC1X 8RR
AE Amsterdam	USA	UK	UK
The Netherlands			

First edition 2006

ISBN-13: 978 0080 445533
ISBN-10: 0080 445535

Printed in Great Britain.

06 07 08 09 10   10 9 8 7 6 5 4 3 2 1

Preface

The aim of this book is to outline the basic physical concepts and device applications related to nanoscience and nanotechnology in semiconductor materials. As illustrated in the book, when the dimensions of a solid are reduced to the size of the characteristic lengths of electrons in the material (de Broglie wavelength, coherence length, localization length, etc.), new physical properties due to quantum effects become apparent. These novel properties are manifested in various ways: quantum conductance oscillations, quantum Hall effects, resonant tunnelling, single-electron transport, etc. They can be observed in properly built nanostructures, such as semiconductor heterojunctions, quantum wells, superlattices, etc. which are described in detail in the text. The effects shown by these quantum structures are not only significant from a purely scientific point of view – several Nobel prices were awarded during the last decades to their discoverers – but also have important practical applications in most of last generation microelectronic and optoelectronic devices.

Only about three decades have elapsed since the pioneering work of Esaki, Tsu, and Chang at the beginning of the 1970s at IBM that established the bases for many of the new effects later observed in quantum wells and superlattices. In order to observe these effects, sophisticated techniques such as molecular beam epitaxy, allowing layer-by-layer growth, and doping of semiconductor nanostructures were routinely set up in many advanced research laboratories during the 1980s. Since all these new developments took place in a relatively short time, it has been difficult to timely incorporate them into the university curricula. However, recently most leading universities have updated their curricula and are offering, both at the graduate and undergraduate levels, courses such as: nanoscale science and engineering, nanoscale structures and devices, quantum devices and nanostructures, etc. Even Masters Degrees are being offered in nanoscale science and engineering. Frequently these courses and titles are included in the schools of physics, materials science, and various engineering schools (electrical, materials, etc.).

In our opinion, there is a lack of comprehensive textbooks at the general undergraduate level dealing with nanoscience and nanotechnology. A few general texts on solid state physics are starting to include several sections, or in some cases, one full chapter, on nanoscale science. Frequently, this material has been added as the last chapter of the new editions of these well-known texts, sometimes without really integrating it in the rest of the book. However, the situation is better for specialized books which can be partially used in graduate courses, since in the last fifteen years a series of excellent texts dealing with nanotechnology and low-dimensional semiconductors have been published. A full reference of these texts is given in the "further reading" sections after the corresponding chapters of our book. They include, among others, the texts of Weisbuch and Vinter (1991), Grahn (1995), Kelly (1995), Ferry and Goodnick (1997), Davies (1999), Mitin, Kochelap, and Stroscio (1999), and Bimberg, Grundmann, and Ledenstov (2001).

Our book is mainly addressed to the final year undergraduate and beginning graduate students in physics, materials science, and several kinds of engineerings (electrical, materials, etc.) with the objective in mind that it could be used in one semester. Alternatively, the book can be of interest to scientists and practicing engineers who want to know about the fundamental aspects of nanoscience and nanotechnology. Our intention has been to write an introductory book on nanoscience and nanotechnology that starting with the physics of low-dimensional semiconductors and quantum heterojunctions would build up to the treatment of those new electronic, transport, and optical properties, which arise as a consequence of both energy quantization of electrons in potential wells and the reduced dimensionality (2D, 1D, 0D) of nanostructures. This process is sequentially carried out showing that the physical concepts involved can be understood in terms of quantum-mechanical and statistical physics theories at the level being taught in the undergraduate school. This is therefore the only real prerequisite for readers to know in advance. We hope that we have succeeded in our aim to show that the concepts related to the already mature field of nanoscience are not more difficult to grasp (with the exception perhaps of the quantum Hall effects) than those corresponding to bulk solid state and semiconductor physics. Once the basic concepts of quantum nanostructures are presented in a unifying scheme, the last chapters of the text deal with applications of nanotechnologies in microelectronics and optoelectronics.

In the process of writing this book, we have always taken into consideration the main objectives already mentioned. Other concerns were the following: (i) The extension of the book has been limited so that it could be taught in one semester. This is especially the case for students having a good knowledge of bulk semiconductor solid state physics, since Chapters 2 and 3 can then be omitted. In addition, some relevant topics in nanotechnology, like carbon nanotubes or biomolecular structures, have been omitted. (ii) We have tried to make as clear as we could the new physical concepts and properties presented in the book, although taking care not to lose the necessary scientific rigour. We have done the same with the mathematical derivations which have been kept as simple as possible. In those cases implying a lot of calculus we have quoted the result, giving the corresponding reference, or have quoted the results of similar calculations in solid state physics, such as, in the applications of the Fermi golden rule. (iii) Due to the introductory character and academic orientation of the book, the bibliography presented at the end of each chapter has been kept to a minimum. Anyway, credit is always given to those scientists who discovered the new phenomena presented in the book or formulated the most admitted theories for their explanation. (iv) To test the understanding of new concepts, a set of exercises is included at the end of each chapter. We have tried that the exercises have a medium degree of difficulty and therefore have avoided those related to extending the theory presented in the text to more difficult cases. We have also given numerous hints directing the students on how to solve the exercises. It is important that the students get the correct final numerical results so that they get an idea of the approximate values of the physical magnitudes involved in nanotechnologies.

About the Authors

J.M. Martínez-Duart obtained his Masters Degree in Physics at the University of Michigan and his PhD both at The Pennsylvania State University and Madrid University. He is the author of about three hundred publications in reputed scientific journals. During the 1970s, he was Assistant Professor at Penn State University and Rensselaer Polytechnic Institute, and Research Visiting Scientist at the IBM T. J. Watson Research Center. Later he has been the Director of the Solid State Physics Institute (CSIC) at Madrid, and the Applied Physics Department at Universidad Autónoma (Madrid). He has been the President of the European Materials Research Society during the period 2000–2001. During the last fifteen years he has been working on the electronic and optoelectronic properties of nanostructured materials. In the field of nanotechnology, Professor Martínez-Duart has been director of several courses, co-author of several books, and Invited Editor of two volumes, "Materials and Processes for Submicron Technologies" and "Current Trends in Nanotechnologies", both published by Elsevier in 1999 and 2001, respectively.

Raúl J. Martín-Palma received the Masters Degree in Applied Physics in 1995 and the PhD in Physics in 2000, both from the Universidad Autónoma de Madrid (Spain). He has been Post-Doctoral Fellow at the New Jersey Institute of Technology (Newark, USA). In spite of his youth, he is the author of over fifty research publications, most of them on electrical and optoelectronic properties of nanostructured materials, published in high impact factor journals. He has been invited to give seminars and conferences at several international scientific congresses. He has received several awards for young scientists for his research on nanostructured materials from the Materials Research Society (USA), European Materials Research Society, and Spanish Society of Materials. He is currently Professor of Physics at the Universidad Autónoma de Madrid.

F. Agulló-Rueda is currently a researcher at the Materials Science Institute of Madrid, which belongs to the Spanish National Research Council (CSIC). He spent several years at the Max-Planck Institute for Solid State Research and at the IBM T. J. Watson Research Center, where he worked on electro-optic effects in semiconductor superlattices. He has published over seventy papers on the optical properties of materials and semiconductor nanostructures and has contributed to several books.

Acknowledgements

The list of acknowledgements has to be necessarily long since the authors have received a great amount of help from many persons. First of all, they would like to acknowledge those scientists who first introduced and directed them into the research field of nanotechnologies, especially E.E. Méndez, L.L. Chang, L. Esaki, H.T. Grahn, R.A. Levy, K.H. Ploog, and K. von Klitzing. We are also much indebted to Professors and graduate students of the Department of Applied Physics (Universidad Autónoma de Madrid) who read various parts of the book and made very helpful suggestions on how it could be improved.

Several chapters of the book have evolved from different courses taught by the authors at Universidad Autónoma de Madrid, in particular, an undergraduate course on electronic and optoelectronic devices based on nanostructures, and a doctoral course on nanostructured materials. In addition, some material has been taken from a short course on nanotechnologies directed by one of us, in parallel with the 2004 European Materials Research Society Conference at Strasbourgh. The text has incorporated many of the suggestions coming from the students attending all these courses as well as from the speakers participating in them.

We would also like to acknowledge L. Dickinson (past, at Elsevier), M. Pacios (European Community, Brussels), and C. Bousuño (Universidad Autónoma de Madrid) for their help in improving the English style of the manuscript as well as in the preparation of some of the figures. We would also like to express our appreciation to Professor H. Grimmeis and other members of the executive committee of the European Materials Research Society for their encouragement in the preparation of this text.

J.M. Martínez-Duart
R.J. Martín-Palma
F. Agulló-Rueda

Structure of the Book

The book comprises ten chapters, which can conceptually be divided into four parts. The first part includes Chapters 1 to 3. Chapter 1 starts by reviewing the present trends in microelectronic and optoelectronic devices. This chapter introduces the reader with the physical bases of semiconductors of reduced dimensionality and with the concepts needed for the definition of nanostructures. Chapters 2 and 3 comprise a short review of solid state and semiconductor physics. These two chapters will be especially useful to non-physics students who have not usually studied these matters with enough rigour and breadth. It is also useful that students are provided with a review of the solid state physics needed for the understanding of nanotechnologies, thus avoiding the frequent consultation of other sources.

The objective of the second part (Chapters 4 and 5) is to expose the reader to the physics explaining the behaviour of electrons in nanostructures and relating it to the density of states function and the energy quantification of the electrons in the different potential wells. The quantum nanostructures studied are the ones more often used in research and in technological applications: semiconductor heterojunctions, quantum wells, superlattices, etc.

The third part (Chapters 6 to 8) is devoted to the transport and optical properties of nanostructures. Chapter 6 deals with transport under the action of electric fields which exhibits some very interesting properties such as quantized conductance, Coulomb blockade, resonant tunnelling, etc. some of them with significant applications in devices. If, in addition of the electric field, a magnetic field is applied (Chapter 7), then both the integral and fractional quantum Hall effects are observed. Even if it does not yet exist a theory fully explaining these two effects, they are studied because they constitute one of the most significant discoveries of the last decades in solid state physics. Similar to the transport properties, the optical properties of nanostructures, studied in Chapter 8, also exhibit a wealth of new properties, very often completely different from those shown by bulk semiconductors: tunability of the gap, microwave Bloch emission, optical properties dependent on the nanocrystal size, etc.

With the above background, the students should be well prepared to tackle the last part (Chapters 9 and 10) on advanced semiconductor devices based on nano-structures. Chapter 9 deals with high frequency transistors (high electron mobility, resonant tunnelling, etc.) as well as single-electron transistors. Finally, Chapter 10 is dedicated to advanced optoelectronic and photonic devices based on quantum heterostructures: quantum well and quantum dot lasers, superlattice photodetectors, high-speed optical modulators, etc. After studying these two chapters, the students should appreciate that it is possible to design advanced devices with pre-fixed electronic and optical characteristics, through what is actually known as "band-structure engineering". Another idea we would like the students to get from these two chapters is that nanoscience is part of today's technology. Many of the devices explained, such as MODFET and heterojunction bipolar transistors, quantum well lasers, photodetectors, modulators, etc. are already commercialized and used in a series of consumer electronic and optoelectronic products.

CONTENTS

Chapter 1

Mesoscopic Physics and Nanotechnologies

Chapter 1

Mesoscopic Physics and Nanotechnologies

1.1. OUTLOOK OF THE BOOK

The interest in the study of the physical properties of electronic materials of very small sizes, usually in the nanometre range, resides in various factors. One of them is due to the trends in microelectronic integrated devices, for which smaller sizes imply operation at higher frequencies, higher functionality, lower fabrication costs for a given performance, etc. A second reason, more important from a scientific point of view, is related to the appearance of new fundamental physical effects, such as resonant tunnelling, quantum conductance, Coulomb blockade, Hall quantum effects, etc. In addition, very frequently, these fundamental discoveries are related to practical devices like quantum well lasers, single electron transistors, confined quantum Stark effect optical modulators, etc.

In Chapter 1 of this book, we start by reviewing the present trends of microelectronic and optoelectronic semiconductor devices, which are the basis for the new field of nanoelectronics. After this we will revise a series of concepts of mesoscopic physics, such as characteristic lengths, needed for the definition of nanostructures. We will also set up the physical basis of semiconductors of reduced dimensionality: quantum wells (2D), quantum wires (1D), and quantum dots (0D). We are aware that some of the concepts introduced in this chapter will be difficult for some readers to grasp, especially if confronted to them for the first time. However, we think that overall it is advantageous to get acquainted with them from the very beginning. After this introductory chapter, a survey of the concepts of quantum mechanics, solid state, and semiconductor physics is presented in Chapters 2 and 3. Chapter 4 deals with the physics of low-dimensional semiconductors, i.e. quantum wells, wires, and dots. In Chapter 5, some of the most frequently used quantum heterostructures are revised, as well as superlattices. The effects of electric and magnetic fields on nanostructures are studied in Chapters 6 and 7, giving special emphasis to the quantum conductance and the quantum Hall effect. The rich variety of optical processes in semiconductor nanostructures is treated in Chapter 8, which completes the basic physical properties of the mesoscopic systems. The last two chapters of the book deal with the electronic and optoelectronic semiconductor devices. In Chapter 9, several kinds of high-frequency diodes and transistors, based on resonant tunnelling and single electron effects, are illustrated. Finally, Chapter 10 is dedicated to the optoelectronic and photonic devices which use quantum heterostructures: quantum well lasers, photodetectors, and optical modulators.

3

1.2. TRENDS IN NANOELECTRONICS AND OPTOELECTRONICS

The evolution of microelectronic devices is influenced by factors such as growing demands in memory capacity of integrated circuits, high transmission data speed, optical communications, etc. This requires electronic devices with faster speed operation and smaller size, as is the case of silicon integrated circuits, in particular dynamic random access memories (DRAMs). Figure 1.1 shows the reduction of feature size of metal-oxide-semiconductor (MOS) transistors, as well as the number of bits per chip for the period 1970–2000 [1]. For example, a 256 M-bit DRAM contains about 10^9 transistors with a feature size L close to 100 nm. For structures with these dimensions, transport can still be treated classically, but we are already at the transition regime to quantum transport (Section 1.8). Today it is believed that present silicon technology will evolve towards feature sizes still one order of magnitude lower, i.e. $L \sim 10$ nm; but below this size, transistors based on new concepts like single electron transistors, resonant tunnelling devices, etc. (Chapter 9) will have to be developed. The operation of this new kind of devices has to be described by the concepts of mesoscopic and quantum physics. It is interesting to remark that quantum effects show up in III-V devices for larger feature sizes, as a consequence of the smaller value of the effective mass, and therefore larger value of the de Broglie wavelength (Section 1.3).

In the near future, and due to the growing demands of calculus from industries like communications, information, military, space, etc. microelectronics will be replaced by nanoelectronics since the feature size of electronic devices will be reduced to about 10 nm.

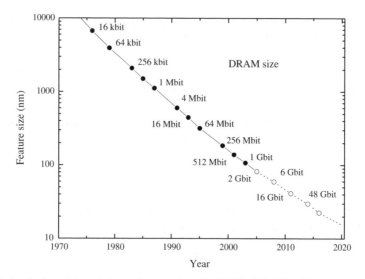

Figure 1.1. Evolution of the minimum feature size of a Si DRAM. After [1].

Although there has been an impressive advance at the device and technology level during the last decade, the progress in the development of new system architectures dealing with billions of transistors is scarce. In this sense, new architectures for parallel data processing are under current development.

The evolution towards nanoelectronics has been possible because of advances in the deposition of very thin films to form heterostructures in which electrons could be confined to a 2D mesoscopic system. Standard evaporation and sputtering techniques did not produce heterostructures of enough quality. However, during the 1980s both molecular beam epitaxy (MBE) and metal organic chemical vapour deposition (MOCVD) became available (Section 1.7). In MBE, a film of the desired material is epitaxially grown over a substrate under ultrahigh vacuum conditions (pressure less than 10^{-10} mbar). The materials are evaporated in tubular effusion cells of the Knudsen type at fairly high temperatures. The atoms or molecules emerge from the effusion cells forming a molecular beam that travels in straight lines towards the substrate where the molecules condense.

Very thin films of elemental semiconductors (Si, Ge), III-V (AlGa, AlGaAs, InP), and II-VI compounds (CdTe, PbS) can be deposited by MBE. This technique allows a layer-by-layer growth of films and superlattices as well as the doping of the material from sources such as B, Al, As, etc. Although the deposited films are of very high quality, the technique is somewhat slow and expensive. MOCVD is the preferred industrial technique to produce semiconductor heterostructures. For instance, GaAs can be grown over a substrate at about $500\,°C$ by the reaction of arsine (AsH_3) gas and trimethyl gallium (CH_3)$_3$Ga at sub-atmospheric pressures. Diluted dopant gases are also simultaneously introduced; for instance Si n-type doping of GaAs can be obtained by means of SiH_4 diluted in hydrogen which is used as carrier gas. This technique allows for simultaneous deposition on several wafers and is used for the commercial production of semiconductor lasers.

In general, mesoscopic systems require the formation of nanostructures in the range close to 100 nm, that is, a decrease in specifications of about one order of magnitude in comparison to the state-of-the-art some 20 years ago. Therefore at present we are close to the limits of conventional optical lithography, and other high resolution nanolithographic techniques (electron beam, ion beam, x-ray, etc.) have to be industrially implemented. As for resists, the most commonly used in nanolithography is the positive tone resist known as PMMA (poly-methylmethacrylate). Although the molecular weight of PMMA is close to 10^6, its roughness, once spin coated onto the substrate, is only about 2 nm.

A frequent matter of present discussion is the ultimate limits in device size, taking into account the evolution shown in Figure 1.1. It seems reasonable that the rate of scaling down predicted by Moore's law [2] will have to slow down. It is expected that the limits of further miniaturization, from an industrial and economic point of view, will be reached in about one decade. The technological limits are related to several factors of which we will mention only two. The first has to do with the amount of heat generated by the power consumed, and which cannot be eliminated because of thermal conductivity limits of the materials and the increasing number of overlayers. The other factor is related to the so-called "parameter spread" in fabrication. For instance many of the electrical parameters of the MOS transistors are set by doping; however, as the size of the region to be doped is decreased to about $0.1\,\mu m^3$, the number of doping atoms becomes so low (about 10) that the parameter spread cannot be controlled appropriately.

In addition to the above mentioned technological limits, there are others of fundamental nature, which are called *physical limits*. Although at present integrated devices are still far from these limits, we think it is important to revise them. They are the following: *(i) Thermal limit*: The energy necessary to write a bit should be several times kT which is the average energy of thermal fluctuations. In CMOS the lowest values contemplated to write a bit should not be smaller than about $2\,eV$, i.e. $100kT$ at room temperature or $\sim 3 \times 10^{-19}$ J. *(ii) Relativistic limit*: Signals cannot propagate faster than the speed of light. Therefore assuming that the nucleus of a microprocessor has a size of a few cm, it takes 10^{-10} s for the signal to propagate, which corresponds to a frequency of about $10\,GHz$. *(iii) Uncertainty principle*: According to the Heisenberg's uncertainty principle, the energy and time needed to write or read a bit should be related by $\Delta E \cdot \Delta t \geq h$. To be safe, we ask the product $\Delta E \cdot \Delta t$ to be $100h$. Since for future circuits ΔE could be as low as 10^{-19} J, we can appreciate that we can approach the quantum limit as the frequency increases.

At present nanoelectronics is moving simultaneously along several directions. One of them is solid state *nanoelectronics*, which is the object of this text, and usually consists of heterostructures of well-known materials (Si, SiO_2, III-V compounds), and several types of transistors: heterojunction, single-electron, resonant-tunnelling, ballistic, etc. However, the amount of computational capacity for some tasks like speech and visual recognition is so large that other radically different alternatives are being sought. Some of these alternatives, like superconductivity electronics and spintronics, use fabrication techniques not too different from those employed in present integrated circuit technologies. *Superconducting electronics*, proposed in the 1970s, and developed to the prototype stage during the 1980s, is based on the switching properties of the Josephson junction, which consists of two superconducting layers separated by a very thin oxide insulating film that can be tunnelled by superconducting pairs of electrons. The advantages of superconducting electronics are based on the fact that Josephson junctions can operate at high switching speeds (switching times between 1 and 10 ps), the amount of dissipated power is very low and the resistance of interconnect superconducting lines is practically null.

Another technology being pursued is *spintronics* which exploits the spin orientation of electrons. Electron-spin transistors are built by enclosing a semiconductor layer (base) sandwiched between two ferromagnetic layers (emitter and collector). Electrons acquiring the magnetization state of the emitter can only travel through the collector across the base if their spins are aligned with the magnetization of the collector. These developments can be considered as running parallel to the efforts in magnetoelectronics to develop MRAM memories, based both in the giant magnetoresistance effect and in the magnetic tunnel junctions developed in 1995. Electron-spin transistors have a big potential if successfully integrated with CMOS circuits.

Other radically new alternatives for future nanoelectronics have been proposed. *Molecular electronics* is based on the different states or configurations that molecules can take, like *"trans"* or *"cis"*, as well as parallel or antiparallel alignment of the spin of unpaired electrons. The change between states must be fast, consume little energy, should be addressed by some external signal, and should be readable by a probe. If this technology is able to be put to work, we will have the ultimate step in miniaturization, since molecules are much smaller than present feature sizes in integrated circuits. Besides, molecules have the advantage of being capable of self-organization in 3D supramolecular entities, although, after the development of scanning atomic force microscopy, molecules can be, in principle, individually manipulated. Examples of molecules that can be used in molecular electronics are: azobenzene, hydrazobenzene, etc. One advantage of organic molecules, when compared to inorganic ones, is that it is easier to isolate them in single molecular systems. Although in the future it is expected that molecular wires or nanotubes can be developed for contacting molecules, today only metallic or semiconducting electrodes are contemplated. Even with this limitation, the interfacing of molecules to the external world for addressability in the case of large systems seems, at present, an enormous problem.

Lastly, we mention biology-inspired electronics, also called *bioelectronics*. In trying to copy nature, we are not concerned with the size of the building blocks, since for instance, a neuron is very large for nanotechnology standards. What nanoelectronics would like to imitate of biological neurons is the capabilities in parallel processing as well as their 3D architectures and the topology of the interconnects. This results from the large number of computations, needed for instance in pattern recognition of the visual systems of humans and animals, which has to be performed simultaneously at different sites. In addition to parallel processing, neural networks try to simulate the integration of computing and memory functions, which in CMOS microprocessors are performed separately.

To conclude this section, the present situation in *optoelectronics* will be shortly considered. Optoelectronic devices, based mainly in III-V direct gap semiconductors, have received a great upsurge since the development of optical fibre communications. In addition, there is at present a tendency to replace, whenever possible, electronic devices by photonic ones. The III-V semiconductors more frequently used are based on AlGaAs–GaAs and GaInAsP–InP heterostructures which cover the 0.8–1.6 μm wavelength range. GaN blue lasers for short wavelength applications were developed about ten years ago. In the last two decades, quantum well semiconductor lasers with very low threshold currents and photodetectors are replacing the conventional ones, especially in long distance optical communications. One very interesting type of quantum well lasers, which operate at still lower threshold currents, is based on strained-layer quantum heterostructures.

At present, laser diodes are manufactured in chips by standard integrated circuit technology, coupled to transistors and optical interconnects, constituting the so-called *optoelectronic integrated circuits (OEIC)*. In all cases, the trend in optoelectronic devices is to achieve a high level of integration which implies smaller sizes, but still in the micron range. Here again there is a lot of research in efficient integration architectures.

Perhaps, the greatest advances in optoelectronics based on quantum semiconductor heterostructures can be found in the field of electro-optical signal modulation. In effect, modulators based on the confined quantum Stark effect (Section 8.4) are several orders of magnitude more effective than their bulk counterparts. As we will see in Section 10.8, this is due to the fact that excitons in quantum wells have much higher ionization energy than in the bulk, and therefore can sustain much higher electric fields.

1.3. CHARACTERISTIC LENGTHS IN MESOSCOPIC SYSTEMS

Mesoscopic physics deals with structures which have a size between the macroscopic world and the microscopic or atomic one. These structures are also called *mesoscopic systems,* or *nanostructures* in a more colloquial way since their size usually ranges from a few nanometres to about 100 nm. The electrons in such mesoscopic systems show their wavelike properties and therefore their behaviour is markedly dependent on the geometry of the samples. In this case, the states of the electrons are wavelike and somewhat similar to electromagnetic radiation in waveguides.

For the description of the behaviour of electrons in solids it is very convenient to define a series of characteristic lengths. If the dimensions of the solid in which the electron is embedded is of the order of, or smaller than these *characteristic lengths,* the material might show new properties, which in general are more interesting than the corresponding ones in macroscopic materials. In fact, the physics needed to explain these new properties is based on quantum mechanics. On the contrary, a mesoscopic system approaches its macroscopic limit if its size is several times its characteristic length.

Let us next describe some of the most commonly used characteristics lengths in mesoscopic systems.

(i) *de Broglie wavelength*

It is well known from quantum mechanics that for an electron of momentum p, there corresponds a wave of wavelength given by the *de Broglie wavelength*:

$$\lambda_B = \frac{h}{p} = \frac{h}{m^* v} \tag{1.1}$$

In Eq. (1.1) we have substituted p by m^*v in a semiclassical description, where m^* is the electron effective mass. From solid state physics, we know that inside a semiconductor, electrons behave dynamically as if their mass was m^*, instead of the mass m_0 of the electron in vacuum. This observation is very important because for many interesting semiconductors, like GaAs or InSb, m^* is much smaller than m_0. For instance, for GaAs and InSb, m^* is equal to $0.067m_0$ and $0.014m_0$, respectively (the concept of effective mass is reviewed in Section 2.6.2). It can be observed, therefore, that the smaller the value of m^*, the easier will be to observe the size quantum effects in nanostructures of a given size. This is the case of semiconductors in comparison with most metals, for which the conduction electrons behave as quasi-free. In fact, with present lithographic techniques, it is relatively easy to construct semiconductor nanostructures with one or two of their dimensions of the order of, or smaller than λ_B.

(ii) *Mean free path*

As the electron moves inside a solid, it is usually scattered by interactions with crystal imperfections like impurities, defects, lattice vibrations (phonons), etc. In most cases, these scattering events or "collisions" are inelastic, i.e. the values of energy and momentum of the system after the interaction, differ from the corresponding ones before they interact. The distance covered by the electron between two inelastic collisions is usually called the *mean free path* ℓ_e of the electron in the solid. If v is the speed of the electron, then

$$\ell_e = v\tau_e \tag{1.2}$$

where τ_e is known as the *relaxation time.*

(iii) *Diffusion length*

In a mesoscopic system of typical size L, the electrons can move either in the ballistic regime or in the diffusive regime. If the previously defined mean free path ℓ_e is much larger than L, the particle moves throughout the structure without scattering; this is the so-called *ballistic transport* regime in which the surfaces usually are the main scattering entities. In hot electron transistors (Section 9.5), electron transport is ballistic and the electrons can reach energies much higher than the ones corresponding to the lattice thermal energy. On the other hand, if $\ell_e \ll L$, transport can be explained as a *diffusion process*. In this case, the system is characterized by a diffusion coefficient D. In terms of D, the *diffusion length* L_e is defined by (see also Eq. (3.51))

$$L_e = (D\tau_e)^{1/2} \tag{1.3}$$

where τ_e is the relaxation time. In semiconductor theory, the concept of diffusion length is used very often; for instance, if electrons diffuse within a p-type semiconductor, their concentration diminishes exponentially with distance with a decay length equal to L_e.

In the diffusive regime, transport in the mesoscopic systems is usually explained by means of the Boltzmann equation, as in the bulk. On the contrary, in the ballistic regime, the Boltzmann transport model is not valid, and electrons move through the structure essentially without scattering.

(iv) *Screening length*

In extrinsic semiconductors, the dopants or impurities are usually ionized and constitute a main factor contributing to scattering. However, in general we cannot consider that the electrical potential produced by these impurities varies with distance as $1/r$. Because of the screening of free carriers by charges of the opposite polarity, the effect of the impurity over the distance is partially reduced. It is found (see, for instance, Ref. [3]) that the variation of the potential is modulated by the term $\exp(-r/\lambda_s)$ where λ_s is called the *screening length* and is given by

$$\lambda_s = \left(\frac{\epsilon kT}{e^2 n} \right)^{1/2} \tag{1.4}$$

where e is the electronic charge, ϵ the dielectric constant of the semiconductor, and n the mean background carrier concentration. One should be careful about nomenclature because some authors call λ_s the Debye length or the Fermi–Thomas length. In a typical semiconductor, λ_s is in the range 10–100 nm, and is an indication of the attenuation of charge disturbances in a semiconductor. From Eq. (1.4) it is determined that λ_s should be much smaller in metals than for semiconductors.

Figure 1.2 shows both the Coulomb potential and the screened potential which varies as

$$\phi_{sp} = \gamma \frac{e^{-r/\lambda_s}}{r} \tag{1.5}$$

where $\gamma = 1/4\pi\epsilon_0$, since for $\lambda_s \rightarrow \infty$ the screening effect disappears, yielding the Coulombic potential. It can be observed from Figure 1.2 that for distances from the impurity larger than about $2\lambda_s$, the screening is almost complete. In the above discussion, we have considered that the charge originating the potential was an impurity, but in general it can be any perturbation in the charge concentration. Another interesting observation is that the above equations imply a dielectric function $\epsilon(r)$ which is distance dependent.

(v) *Localization length*

The localization length can be understood in terms of transport in disordered materials, in which we know from solid state physics that, in addition to Bloch extended

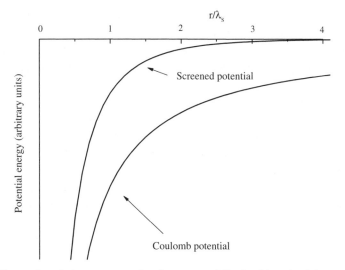

Figure 1.2. Comparison between screened and unscreened Coulombic potential.

states, there can also be localized states (see also Section 7.7.3). In disordered materials, the electrons move by "hopping" transport between localized states, or from a localized state to a bound state.

In order to describe the hopping transport and other mesoscopic properties of the localized states, it is assumed that the electron wave function is described by

$$\psi = \exp(-r/\lambda_{loc}) \tag{1.6}$$

where λ_{loc} is known as the *localization length*. Evidently, the electrical conductivity of a material will be proportional, among other factors, to the overlap between the wave functions. If the sample dimensions are of the order of λ_{loc}, we can say that our system is mesoscopic. In fact, we will use the concept of localized states to explain the quantum Hall effect in Chapter 7. The type of localization just mentioned is also similar to the Anderson localization used to explain the metal–insulator transition in solid state physics [4].

1.4. QUANTUM MECHANICAL COHERENCE

In a mesoscopic structure of dimensions similar to the electron de Broglie wavelength λ_B, the behaviour of the electron should be described quantum mechanically, i.e. by using Schrödinger equation. If the electron interacts inelastically with a defect, or any impurity, the electrons change their energy and momentum, as well as the phase of their wave function. The *phase coherence length* L_ϕ is defined as the distance travelled by the electron without the carrier wave changing its phase. Evidently, interference effects in the electron waves should only be observed if the particles move over distances of the order of, or smaller than, L_ϕ.

Since we are usually interested in mesoscopic systems in the quasi-ballistic regime, in which electrons are practically unscattered, L_ϕ should be a length similar to the inelastic scattering mean free path ℓ_e previously defined. Coherent states can evidently show interference effects. On the contrary, once the coherent states loose their coherence, by inelastic scattering, the corresponding waves cannot be superposed and cannot interfere (in some way, the matter–wave becomes a particle). In mesoscopic physics, the loss of coherence is usually called *dephasing*. Evidently, the coherence processes are characteristic of mesoscopic systems.

From the above definition of L_ϕ, electrons can show *interference effects* over distances smaller than L_ϕ. If electrons with phase ϕ_1 interfere with electrons of phase ϕ_2, we know from the wave theory that the amplitude of the resultant wave varies as $\cos(\phi_1 - \phi_2)$ and the amplitudes can add up to each other, or they can be subtracted depending on the phase difference. Interference effects in mesoscopic systems will be studied in Chapter 7. For instance, in the Aharonov–Bohm effect (Section 7.5) we will see how magnetic fields (in reality, the vector potential) modulate the phase difference between two electron currents which are added after having travelled through different channels. We will see that variations of magnetic flux modulate the conductance (or its inverse, the resistance) of the mesoscopic system in quantum units.

1.5. QUANTUM WELLS, WIRES, AND DOTS

In Section 1.3, we have defined a series of characteristic lengths λ which correspond to physical properties of electrons which are size dependent. We have also seen that when the dimensions of the solid get reduced to a size comparable with, or smaller than λ, then the particles behave wave-like and quantum mechanics should be used.

Let us suppose that we have an electron confined within a box of dimensions L_x, L_y, L_z. If the characteristic length is λ, we can have the following situations:

(i) $\lambda \ll L_x, L_y, L_z$

In this case the electron behaves as in a regular 3D bulk semiconductor.

(ii) $\lambda > L_x$ and $L_x \ll L_y, L_z$

In this situation we have a 2D semiconductor perpendicular to the *x*-axis. This mesoscopic system is also called a *quantum well* (see, for instance, Figure 4.1, Chapter 4).

(iii) $\lambda > L_x, L_y$ and $L_x, L_y \ll L_z$

corresponds to a 1D semiconductor or *quantum wire*, located along the z-axis (Section 4.5).

(iv) $\lambda \gg L_x, L_y, L_z$

In this case it is said that we have a 0D semiconductor or a *quantum dot* (Section 4.6).

In general, we say in mesoscopic physics that a solid, very often a semiconductor, is of *reduced dimensionality* if at least one of its dimensions L_i is smaller than the characteristic length. For instance, if L_x and L_y are smaller than λ we have a semiconductor of dimensionality equal to one. We could also have the case that λ is comparable, or a little larger, than one of the dimensions of the solid but much smaller than the other two. Then we have a *quasi* 2D system, which in practice is a very thin film, but not thin enough to show quantum size effects.

1.6. DENSITY OF STATES AND DIMENSIONALITY

Although the *density of states* (DOS) of physical systems will be derived formally in Chapter 4, in this section we see from a mathematical point of view the consequences of the dimensionality of the system in the DOS. As we know from solid state physics, most physical properties significantly depend on the DOS function ρ. The DOS function, at a given value E of energy, is defined such that $\rho(E)\Delta E$ is equal to the number of states (i.e. solutions of Schrödinger equation) in the interval energy ΔE around E. We also know that if the dimensions $L_i (i = x, y, z)$ are macroscopic and if proper boundary conditions are chosen, the energy levels can be treated as a quasi-continuous. On the other hand, in the case where any of the dimensions L_i gets small enough, the DOS function becomes discontinuous. Let us next obtain the DOS function for several low-dimensional solids. First, let us remind that for bulk solids $\rho(E)$ varies with energy in the form $\sqrt{E}$ (Section 2.3).

If each quantum state or Bloch state (Sections 2.4 and 2.5) in a solid is designated by a quantum number k (Bloch state), the general expression for the DOS function should be

$$\rho(E) = \sum_k \delta(E - E_k) \tag{1.7}$$

where the quantized energies are given by E_k and $\delta(E)$ is the Dirac's delta function. If we take into account the electron spin degeneracy, a factor 2 should also appear in the above expression. Let us recall for simplicity the case of a cubic shaped 3D macroscopic crystalline solid, of edge $L = Na$, where a is the lattice constant and N the number of sites along the one-dimensional directions which is supposed to be large. In this case, the eigenstates can be considered as quasi-continuous and the summatory in k of Eq. (1.7) can be replaced by an integral, i.e.

$$\sum_{k} \rightarrow \frac{L}{(2\pi)^3} \int dk \qquad (1.8)$$

for the case of a cube of size L and volume $V = L^3$ (Section 2.3). We also know from simple solid state theory that if we assume that the energy E_k only depends on the magnitude of k in a parabolic energy dependence between the momentum $\hbar k$, i.e. $E_k = \hbar^2 k^2 / 2m^*$, then (Section 2.3):

$$\rho_{3D}(E) = \frac{V}{2\pi^2} \left(\frac{2m^*}{\hbar^2} \right)^{3/2} \sqrt{E} \qquad (1.9)$$

We can follow exactly the same procedure for 2D and 1D semiconductors of area A and length L, respectively (see Sections 4.2 and 4.5), reaching the following expressions for $\rho_{2D}(E)$ and $\rho_{1D}(E)$:

$$\rho_{2D}(E) = \frac{A}{\pi} \left(\frac{m^*}{\hbar^2} \right) \qquad (1.10)$$

$$\rho_{1D}(E) = \frac{L}{2\pi} \left(\frac{2m^*}{\hbar^2} \right)^{1/2} \frac{1}{\sqrt{E}} \qquad (1.11)$$

Some important considerations can already be made: the DOS function in 3D semiconductors is proportional to $\sqrt{E}$, in 2D is constant, and in 1D varies inversely proportional to $\sqrt{E}$. This implies in the last case that at the bottom of bands, the DOS plays a very important role, because there is a singularity for $E = 0$.

Eqs (1.10) and (1.11) were derived for perfectly 2D and 1D solids, but in the real world a 2D solid, for instance, is really a 3D one where the perpendicular dimension is very short. We will see in Section 4.2 that since electrons can move almost freely in the (x, y) plane, Eq. (1.10) should be written for a quasi-2D solid as

$$\rho_{2D}(E) = \frac{A}{\pi} \left(\frac{m^*}{\hbar^2} \right) \sum_{n_z} \theta(E - E_{n_z}) \qquad (1.12)$$

where n_z refers to the quantization in the confined z-axis and θ is the step function. This function is represented in Figure 4.3.

Similarly, for a quasi-1D solid or quantum wire along the z-direction

$$\rho_{1D}(E) = \frac{L}{2\pi} \left(\frac{2m^*}{\hbar^2} \right)^{1/2} \sum \frac{1}{\sqrt{E - E_{n_x, n_y}}} \tag{1.13}$$

where n_x and n_y are the quantum numbers for the confined x and y directions. This function looks like a series of peaks, similar to the one represented in Figure 4.7, one for each value of E_{n_x, n_y}.

Evidently, in the case of a quasi-0D solid or quantum dot, there is no continuous DOS function, since there is quantization in the three spatial directions. Therefore the DOS function consists of a series of peaks given by

$$\rho_{0D}(E) = \sum_i \delta(E - E_i) \tag{1.14}$$

where $i = (n_x, n_y, n_z)$ (see Figure 4.8). In reality, the peaks in quantum dots are not perfect δ-functions since there is a broadening effect as a consequence of scattering mechanisms.

1.7. SEMICONDUCTOR HETEROSTRUCTURES

Although a very large number of semiconductor electronic devices (of the order of 100) can be individually identified, they can be considered as being built from only a few fundamental structures. These structures are: (i) The p–n homojunction, mainly based on crystalline silicon; (ii) the metal–semiconductor interface; (iii) the metal–insulator–semiconductor structure, in particular the Si–SiO$_2$–metal or MOS structure; and (iv) the semiconductor heterojunction or interface between two semiconductors of different gaps. Most of the electronic and optoelectronic devices studied in this text (Chapters 9 and 10) are precisely based on heterojunctions. In addition, heterojunctions are very appropriate for fundamental studies in mesoscopic physics, in particular 2D electron systems in quantum wells. In fact, K. von Klitzing discovered the quantum Hall effect in a commercial MOS structure (Section 7.7) [5].

At present, the best quality quantum heterostructures are made of III-V semiconductors. The one most often used is the AlGaAs–GaAs heterostructure which is treated in detail in Section 5.3. These heterostructures have been routinely produced since about 20 years ago. Figure 1.3 shows the band diagram of the simplest AlGaAs–GaAs heterojunction. As explained in Section 5.3, the most interesting effect of this heterojunction arises from the large concentration of electrons, produced in the n-type doped AlGaAs, which drop

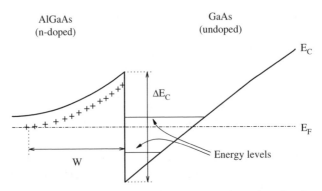

Figure 1.3. Conduction band diagram of a modulation-doped heterojunction (see also Figure 5.3 in Section 5.3.1).

into the quantum well at the interface, and become spatially separated from the impurities. This heterostructure is also known as *modulation-doped* since the doping is selective. If an electric field, as in the MOSFET transistor, is applied parallel to the interface, the electrons can move extremely rapidly, as a consequence of the small effective mass of electrons in GaAs, and more importantly, due to the lack of impurity scattering inside the quantum well. In addition, the electrons move in a quantum well of typical width smaller than λ_B and therefore become quantified in the direction perpendicular to the interface. The electron system in the well has the properties of a 2D system and the DOS is given by Eq. (1.12) which, as we will see in Section 5.3, results in much improved electronic and optical properties.

Modulation-doped heterostructures of high quality could be fabricated only after such techniques as molecular beam epitaxy and metal organic chemical vapour deposition were fully developed. These techniques allow to deposit very thin films, which grow atomic layer by layer on a high quality substrate under a vacuum of about 10^{-12} mbar, resulting in highly regular interfaces (no roughness) in the planar directions and very abrupt composition gradients in the perpendicular direction. The quality of this interface is much higher than the Si–SiO$_2$ interface of MOS devices (Section 5.2) and therefore allows a better observation of quantum effects.

Semiconductor quantum heterojunctions are at present the base of many electronic and optoelectronic commercial devices. In this sense, we can say that some of the most advanced electronic and optoelectronic devices are based on nanotechnologies. As we will see in detail in Chapter 9, resonant tunnelling diodes are based on two heterojunctions with a quantum well between them; also the high mobility transistor (Section 9.2) is based on one heterojunction. In the same line, advanced optoelectronic devices like lasers for optical communications, CD players, and very fast electro-optical modulators are based on quantum well heterostructures (Chapter 10).

1.8. QUANTUM TRANSPORT

Transport studies in nanostructures during the last 20 years have been mainly focused in the study of quantum interference effects (Section 1.4) and in single-electron transport (Section 6.4.3). Since the electron mean free path is usually much larger in semiconductors than in metals, interference effects are better observed in semiconductor nanostructures. Most of the earlier studies in macroscopic transport distinguished between the transport in the *diffusive regime* and transport in the *ballistic regime* (Section 1.3(iii)). Diffusive transport is practically independent of the shape of the system; therefore, the electron numerous scattering mechanisms are practically similar to those in bulk materials. On the other hand, in quantum heterostructures for which $\ell \gg L$ the electron travels ballistically and only interacts with the system boundaries. In this case, in which in addition λ_B is comparable or larger than L, the energy quantization of the electrons in the well becomes very significant. Ballistic electrons that travel through the structure without scattering, and therefore, can show interference effects. Frequently, these effects, treated in Chapter 7, are observed under the action of magnetic fields.

In metallic nanostructures, ℓ is of the order of $100\,\text{Å}$ and one usually has a diffusive regime. On the other hand, in semiconductor heterostructures ℓ is often of the size of several micrometers, and the confinement effects become much more significant. In this situation, we cannot describe transport by making use of macroscopic properties, like electrical conductivity, and have to appeal to the wavelike properties of the wave function given by Schrödinger equation. In fact, a new formulation, due to Landauer and Büttiker, will have to be used, which will be able to explain the observed quantization of conductance (Section 6.4). *Quantum transport* also deals with electron tunnelling across potential barriers of controlled height and width. One example is the resonant tunnelling diode, which shows very interesting effects like a tunnelling probability of practically 100% for certain energies as well as a negative differential resistance (see Section 9.4) in their *I–V* characteristics.

REFERENCES

[1] Mitin, V.V., Kochelap, V.A. & Stroscio, M.A. (1999) *Quantum Heterostructures* (Cambridge University Press, Cambridge).

[2] Goser, K., Glösekötter, P. & Dienstuhl, J. (2004) *Nanoelectronics and Nanosystems* (Springer, Berlin).

[3] Kittel, Ch. (2005) *Introduction to Solid State Physics*, 8th edition (Wiley, New York).

[4] Anderson, P.W. (1958) *Phys. Rev.*, **109**, 1492.

[5] von Klitzing, K., Dorda, G. & Pepper, M. (1980) *Phys. Rev. Lett.*, **45**, 494.

FURTHER READING

Goser, K., Glösekötter, P. & Dienstuhl, J. (2004) *Nanoelectronics and Nanosystems* (Springer, Berlin).

Imry, Y. (1997) *Introduction to Mesoscopic Physics* (Oxford University Press, Oxford).

Mesoscopic Physics and Electronics (1998) Eds. Ando, T., Arakawa, Y., Furuya, K., Komiyama, S. and Nakashima, H. (Springer, Berlin).

Murayama, Y. (2001) *Mesoscopic Systems* (Wiley-VCH, Weinheim).

PROBLEMS

1. **Landau levels in a bulk solid**. (The purpose of this exercise is to remind the reader that low-dimensional confinement effects can also appear in bulk solids under the action of a magnetic field). Show that the energy levels of a 3D solid in a magnetic field B along the z-direction is given by

$$E(n, k_z) = E(n) + E(k_z) = \left(n + \frac{1}{2}\right) \hbar w_c + \frac{\hbar^2 k_z^2}{2m^*}, \quad n = 0, 1, 2, K$$

where the cyclotron frequency w_c is given by $w_c = eB/m^*$ (we have assumed that the constant energy surfaces in k-space are spherical so that the effective mass is a scalar).

2. **Magnetic length**. The characteristic magnetic length ℓ_m for electrons subjected to the action of a magnetic field B is defined as

$$\ell_m = \left(\frac{\hbar}{eB}\right)^{1/2}$$

Show that ℓ_m is equal to the radius of the electron cyclotron orbit corresponding to the value of the quantized energy in the plane perpendicular to B (see problem 1.1) for $n = 0$. *Hint*: you can use a semiclassical argument and therefore write $E = (1/2) \, m^* v^2$ with $v = r w_c$, w_c being the cyclotron angular frequency.

3. **Coherence length in superconductors**. Compare the coherence length ξ of a superconductor with the characteristic lengths defined in Section 1.3. *Hint*: the definition of ξ comes out in a natural way from the phenomenological Ginzburg–Landau theory of superconductivity. For very pure superconductors, the mean free path ℓ in the normal state is quite large and it is shown that $\xi = \hbar v_F / \pi E_g$ where v_F is the Fermi velocity and E_g the superconductor gap. However, for very impure samples ℓ is short and $\xi = (2\hbar v_F \ell / \pi E_g)^{1/2}$.

4. **Density of states function for a solid in a magnetic field**. From the expression of $E(n, k_z)$ of problem 1, show that the density of states of a bulk solid in a magnetic field as a function of energy is equal to a series of infinite one-sided hyperbolic step functions similar to Figure 4.7. In addition, show that analytically it can be written in the form:

$$ n_{1D} \propto \sum_{n=0}^{\infty} (E - E_n)^{-1/2} $$

where E_n is given by the quantized energies of problem 1. Represent n_{1D} as a function of E showing that has singularities ($n_{1D} \rightarrow \infty$) for values of the energy given by $E_n = \hbar w_c/2, 3\hbar w_c/2$, etc.

5. **Resistivity and conductivity tensors in 2D**. (The expressions for the σ and ρ tensors in 2D are useful for the study of magneto-transport and are used for the interpretation of the quantum Hall effect in Chapter 7). In a 2D system in which the current flows in the x-direction, as in Hall measurements, with the magnetic field applied along the z-direction, show that the resistivity and conductivity tensors are related by the following expressions

$$ \rho_{xx} \equiv \frac{\sigma_{xx}}{\sigma_{xx}^2 + \sigma_{xy}^2} $$

$$ \rho_{xy} \equiv \frac{\sigma_{xy}}{\sigma_{xx}^2 + \sigma_{xy}^2} $$

Hint: assume a relation between current and electric field of the form

$$ J_x = \sigma_{xx} E_x + \sigma_{xy} E_y $$

$$ J_y = \sigma_{yx} E_x + \sigma_{yy} E_y $$

and that the sample is homogeneous and isotropic, i.e. $\sigma_{xx} = \sigma_{yy}$ and $\sigma_{xy} = -\sigma_{yx}$.

6. **Hall coefficient in 2D**. (The solution of this problem gets us used to handle 2D electron systems under the action of a magnetic field). Assuming the geometry of the previous problem, with a magnetic field B in the z-direction, show that: (a) the components of the velocity for the electron motion in the (x, y) plane are given by

$$ v_x = -\frac{e\tau}{m^*} E_x - w_c \tau v_y $$

$$ v_y = -\frac{e\tau}{m^*} E_y + w_c \tau v_x $$

Hint: assume the relaxation time approximation of transport in solids, reviewed in Section 3.5.1.

(b) If $v_y = 0$, as in a Hall measurement show that the Hall coefficient $R_H (\equiv E_y / J_x B)$ is given by

$$R_H = \frac{E_y}{J_x B} = \frac{v_y}{IB} = -\frac{1}{N_s' e}$$

where N_s' is the electron concentration per unit area and $J_x = I/a$, where a is the width of the sample.

Chapter 2

Survey of Solid State Physics

Chapter 2

Survey of Solid State Physics

2.1. INTRODUCTION

Before we study the effects of reduced size and dimensionality on the properties of solids, we review in this chapter those concepts of solid state physics which are essential for the understanding of the behaviour of quantum nanostructures. For instance, the behaviour of electrons in a quantum well is very different to the case of bulk solids if their motion is across the potential barriers confining the quantum well, but is very similar if the motion is parallel to the interfaces. In Section 2.2 of this chapter we review some basic concepts of quantum mechanics, especially the time-independent and time-dependent perturbation techniques which lead to the Golden rule for electron transitions between states. Section 2.3 reviews the electron model of solids following the lines proposed by Sommerfeld, based on the fact that quantum mechanics only allows a certain number of states for a particle in a box. The model ignores the details of the crystalline periodic potential and assumes a mean potential V_0. This model explains qualitatively many properties of solids: magnitude and temperature dependence of electronic heat capacity, the Wiedemann–Franz ratio between thermal and electrical conductivities of metals, etc. However, this model cannot explain other properties like the existence of Hall coefficients with positive sign in some metals, the fact that some materials are semiconductors and other metals, etc.

Section 2.4 deals with Bloch theorem which introduces the formalism for the explanation of the behaviour of electrons in periodic potentials. Subsequently, the Bloch formalism is applied in Section 2.5 to two limiting situations: the case of a weak periodic potential (nearly free electron model) and the case of a potential so strong that in a first approximation assumes that the atomic potentials are the main components to the total energy (tight binding approximation). Surprisingly enough both limits give results which are similar from a qualitative point of view, and that can be summarized in the existence of an electronic band structure, i.e. the electrons can move within the allowed energy bands which are separated by forbidden bandgaps.

Section 2.6 deals with electron dynamics in solids. The electron wave function, as defined by Bloch, leads immediately to the concept of the velocity of the electron (the group velocity of the superposition of Bloch states). When electrons are subjected to external forces, the concept of the effective mass tensor is very useful to characterize the dynamics of the electrons in the band, since it is linked to the curvature of the bands at a given energy. The chapter ends (Sections 2.7 and 2.8) with the study of lattice dynamics based on a description of the atomic vibrational normal modes in a crystal. As electrons in periodic structures, there are only allowed vibrational characteristic frequencies, dependent on the wave vector. Next, in order to make the language "more physical", the concept of phonons is introduced in the study of lattice dynamics. Theoretically, the phonon is a quasi-particle that can interact or suffer collisions with electrons and the concept is very useful for the study of transport properties of solids.

2.2. SHORT REVIEW OF QUANTUM MECHANICS

In this section, we will review some of the basic concepts of quantum mechanics, which are necessary to understand the properties of bulk and reduced dimensionality semiconductors, as well as the operation of modern electronic and optoelectronic devices.

2.2.1. *Wave-particle duality and the Heisenberg principle*

In classical physics we deal with two kinds of entities: particles, such as a small mass which obeys Newton's equations, and waves as, for instance, electromagnetic waves or light which behave according to Maxwell's equations. However, in dealing with very small objects, like atoms, the above classification is not enough to describe their behaviour, and we have to turn to quantum mechanics, and to the dual concept of wave-particle. For instance, if light interacts with a material, it is better to think of it as being constituted by particles called photons instead of waves. On the other hand, electrons of which we have the primary concept of particle, behave like waves when they move inside a solid of nanometre ($= 10^{-9}$ m) dimensions.

When we study in optic courses the phenomena of light interference and diffraction, we assume that light behaves as a wave of wavelength λ and frequency v, related by the expression $c = \lambda v$, where c is the speed of light in the medium of propagation, usually vacuum or a transparent medium. However, Planck in 1901 assumed, in order to explain the radiation from a black body, that light can only be emitted or absorbed in units of energy quanta called *photons*. The energy of the photons is

$$E = hv = \hbar\omega \tag{2.1}$$

where h = Planck's constant ($= 6.62 \times 10^{-34}$ J·s), $\hbar = h/2\pi$, and $\omega = 2\pi v$ is the angular frequency.

The concept of light behaving as photons was completely admitted after Einstein explained in 1905 the photoelectric effect. According to this effect, electrons are emitted from the surface of a material when photons of sufficient energy impinge against it. Evidently, the energy of the impinging photons has to be higher than the energy of the barrier that the electrons inside the solid have to surmount to be emitted. Einstein also showed that a photon of energy E has a momentum given by

$$E = cp \tag{2.2}$$

where c is the speed of light. It is also convenient to remember that since photons move at the speed of light, their mass at rest should be zero.

At first, it was usual to think of electrons as particles, as for instance, when they strike against a phosphorescent screen after being accelerated. However, in 1927, Davison and Germer showed that electrons impinging against a nickel crystal surface were diffracted, as if they were waves, and in fact followed Bragg's law of diffraction. We can appreciate, therefore, that the *concept of wave-particle* is well established by experimental results. In fact, as *de Broglie* vaticinated in 1924, to every particle of momentum p, a wave of wavelength

$$\lambda = \frac{h}{p} \tag{2.3}$$

can be associated, Eq. (2.3) can also be written as:

$$p = \frac{h}{\lambda} = \frac{h}{2\pi} \frac{2\pi}{\lambda} = \hbar k \tag{2.4}$$

where k is the so-called wave number.

One very important principle of quantum mechanics was enunciated by Heisenberg in 1927. According to the *Heisenberg uncertainty principle,* in any experiment, the products of the uncertainties, of the particle momentum Δp_x and its coordinate Δx must be larger than $\hbar/2$, i.e.:

$$\Delta p_x \Delta x \geq \hbar/2 \tag{2.5}$$

There are of course corresponding relations for $\Delta p_y \Delta y$ and $\Delta p_z \Delta z$. It is important to remark that this indeterminacy principle is inherent to nature, and has nothing to do with errors in instruments that would measure p_x and x simultaneously. The second part of this principle is related to the accuracy in the measurement of the energy E and the time interval Δt required for the measurement, establishing that

$$\Delta E \Delta t \geq \frac{\hbar}{2} \tag{2.6}$$

2.2.2. *Schrödinger wave equation. Applications*

As it was established by Schrödinger in 1926, the dual wave-like and particle-like properties of matter are described by the so-called *wave function* $\Phi(\vec{r}, t)$, which is continuous and has continuous derivatives. The wave function is also complex, i.e. it has a real part and an imaginary one. The wave function satisfies the second-order, linear, differential *Schrödinger equation*:

$$\left[-\frac{\hbar^2}{2m} \nabla^2 + V(\vec{r}, t) \right] \Phi = i\hbar \frac{\partial \Phi}{\partial t} \qquad (2.7)$$

where ∇^2 is the operator $\partial^2/\partial x^2 + \partial^2/\partial y^2 + \partial^2/\partial z^2$, and V is the potential energy, which is generally a function of position and possibly of time. Although the function Φ does not have a physical meaning, the product of Φ by its complex conjugate is a real quantity, such that the probability dP of finding a particle in a small volume dV is given by

$$dP = |\Phi^2| dV \qquad (2.8)$$

If the potential energy V is not time dependent, we can search for a solution to Eq. (2.7) of the form

$$\Phi(\vec{r}, t) = \psi(\vec{r}) e^{-i\omega t} \qquad (2.9)$$

Substituting Eq. (2.9) in (2.7) and writing $E = \hbar\omega$, we find the *time-independent Schrödinger equation*

$$\left[-\frac{\hbar^2}{2m} \nabla^2 + V(\vec{r}) \right] \psi(\vec{r}) = E\psi(\vec{r}) \qquad (2.10)$$

for the time-independent wave function $\psi(\vec{r})$.

Schrödinger's equation can be solved exactly only in a few cases. Probably the simplest one is that of the *free particle*, as for instance a free electron of energy E and mass m. In this case $V(\vec{r}) = 0$ and the solution of Eq. (2.7) is easily found to be

$$\Phi = A e^{i(kx - \omega t)} + B e^{i(-kx - \omega t)} \qquad (2.11)$$

where

$$k = \left(\frac{2mE}{\hbar^2} \right)^{1/2} \qquad (2.12)$$

Therefore, the free electron is described by a wave, which according to the de Broglie relation has momentum and energy given, respectively, by

$$p = \hbar k, \qquad E = \frac{p^2}{2m} \tag{2.13}$$

In general we will assume that the electron travels in one direction, for instance, along the x-axis from left to right, and therefore the coefficient B in Eq. (2.11) is zero. The wave equation for the free electron can simply be written as:

$$\Phi = A e^{i(kx - wt)} \tag{2.14}$$

Another example in which Schrödinger's equation can be solved exactly is that of the *hydrogen atom* for which the potential is Coulombic, i.e. V varies with the distance r between the proton and electron in the form $1/r$. Solving Schrödinger's equation, one gets the well-known expression for the electron energies:

$$E_n = -\frac{m_r e^4}{2(4\pi \varepsilon_0)^2 \hbar^2} \frac{1}{n^2} = -\frac{13.6}{n^2} \text{(eV)}, \qquad n = 1, 2, \ldots \tag{2.15}$$

where m_r is the reduced proton–electron mass. In solid state physics, the mathematical model of the hydrogen atom is often used, as for example, in the study of the effects of impurities and excitons in semiconductors. Although the equation giving the values of the energy is very similar to Eq. (2.15), the values of the energy are much smaller, since the dielectric constant of the medium has to substitute the value of the permittivity of vacuum ε_0. For instance, in the case of silicon, the value of the dielectric constant is about $12\varepsilon_0$.

The model of the *harmonic oscillator* is very often used in solid state physics, for instance, in the study of lattice dynamics in solids (Sections 2.7 and 2.8). In general terms, we can say that the harmonic oscillator model can be used to describe any system which performs vibrations of small amplitude about an equilibrium point. The allowed energies of the harmonic oscillator can be obtained from Schrödingers's equation and are given by:

$$E_n = \left(n - \frac{1}{2}\right)\hbar\omega, \qquad n = 1, 2, 3, \ldots \tag{2.16}$$

Another potential that allows the exact solution of Schrödinger's equation is the so-called *infinite square well potential*, which consists of a flat potential of width a surrounded by infinite potentials at its extremes. This potential is considered in detail in Sections 4.2 and 5.4.1, since it is a fairly good approximation to modulation-doped quantum wells, which is the basic structure of many quantum well transistors and lasers.

2.2.3. Fermi–Dirac and Bose–Einstein distributions

In this section we will deal often with distribution functions of several particles of different nature, for instance electrons, photons, etc. Electrons and other particles with half-integer spin are called *fermions* and obey the Pauli exclusion principle and Fermi–Dirac statistics. On the other hand, particles with zero or integer spin obey Bose–Einstein statistics and are called *bosons*.

In equilibrium, the average number of fermions that occupy an energy state E is given by the *Fermi–Dirac distribution function* f_{FD} (E, T):

$$f_{FD}(E, T) = \frac{1}{\exp[(E - E_F)/kT] + 1} \tag{2.17}$$

which gives the probability that a given energy state is occupied. The energy E_F is called the Fermi energy, or *Fermi level* in semiconductors. From Eq. (2.17), we see that f_{FD} is equal to 0.5 when $E = F_F$. Figure 2.1 represents the Fermi–Dirac distribution function at several temperatures. Evidently, all curves have the value 0.5 for $E = E_F$. At $T = 0$ K, f_{FD} is a step function, all states of energy below E_F being occupied and those for energies above E_F being empty. At any temperature, the transition at E_F from 1 to 0 occurs in a relatively small energy width of about $4kT$.

Bosons obey the *Bose–Einstein distribution function* $f_{BE}(E, T)$ which is given by

$$f_{BE}(E, T) = \frac{1}{\exp(E/kT) - 1} \tag{2.18}$$

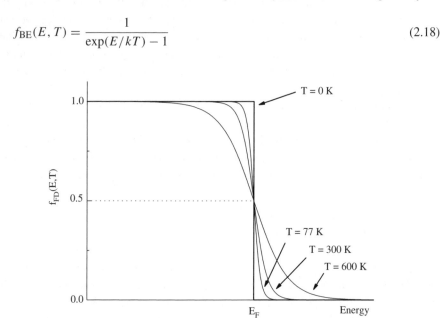

Figure 2.1. Fermi–Dirac distribution function as a function of energy, at different temperatures.

The negative sign of 1 in the denominator, in comparison with the positive sign that appears in Eq. (2.17) for f_{FD}, makes a significant difference. For instance, if $E \to 0$ f_{BE} goes to infinity. For this reason bosons, in particular ^{4}He, undergo Bose condensation, i.e. all bosons "condense" in the ground state. As a contrast, fermions are forced apart, since there can be only one particle in each state.

As it is easily appreciated from Eqs (2.17) and (2.18), for high energies in relation to kT, the 1 that appears in the denominator can be neglected, and both distribution functions approach the classical *Boltzmann distribution function*. Thus, if we have, for instance, electrons in a semiconductor for which $(E - E_F) \gg kT$, then

$$f(E, T) \approx \exp\left(-\frac{E - E_F}{kT}\right) \tag{2.19}$$

Observe that for energies much higher than kT, both quantum distributions converge to the classical one.

2.2.4. Perturbation methods

We have seen in Section 2.2.2 some important problems in quantum mechanics, like the harmonic oscillator, for which Schrödinger equation could be solved exactly. However, in most cases we must deal with potentials for which Schrödinger equation does not provide exact solutions. Probably the most important method for obtaining approximate solutions is the *perturbation method*. This method can be applied whenever the Hamiltonian H of the system is not too different from a Hamiltonian H_0 which describes a system whose eigenfunctions can be found exactly. Therefore we assume that

$$H = H_0 + H' \tag{2.20}$$

where H' is called the perturbation Hamiltonian.

Our aim is to find the eigenvalues E_n and eigenfunctions ψ_n of H, i.e.

$$H\psi_n = E_n\psi_n = (H_0 + H')\psi_n \tag{2.21}$$

We have assumed that the unperturbated eigenvalues ξ_n and eigenfunctions ϕ_n can be exactly calculated by solving the equation

$$H_0\phi_n = \xi_n\phi_n \tag{2.22}$$

The unknown perturbed states ψ_n are expressed as a superposition of the complete set of unperturbed wave functions, i.e.

$$\psi_n = \sum_i c_{in} \phi_i \qquad (2.23)$$

Standard books in quantum mechanics [1] show how to proceed to obtain several approximations to the eigenstates. For the zero-order approximation, we evidently get $E_n = \xi_n$. Very often the *first-order approximation* suffices, and is given by

$$E_n \cong \xi_n + H'_{nn} = \xi_n + < \phi_n | H' | \phi_n > \qquad (2.24)$$

which only involves the zero-order wave functions.

Generally, the unperturbed Hamiltonian H_0 is usually symmetrical and the perturbation Hamiltonian H' has a specific symmetry, either even or odd. In this case, it is easy to prove that for H' even, all matrix elements between states of opposite parity are zero, and, on the contrary, if H' is odd, all matrix elements between states of the same parity automatically vanish. These results are equivalent to the selection rules in spectroscopy.

The perturbation theory discussed so far applies to time-independent stationary states. In order to study the *time-dependent perturbation theory*, suppose next that one initial state is subjected to some time-dependent force. As an example, imagine the oscillation force exerted on an atom by a light wave. As in the previous case we assume that the perturbation term $H'(t)$ in the Hamiltonian is small. We now try to find the eigenvalues and eigenfunctions of the time-dependent Schrödinger equation (see Eq. (2.7))

$$H\psi = [H_0 + H'(t)]\psi = ih\frac{\partial \psi}{\partial t} \qquad (2.25)$$

Here again we can express ψ as a superposition of the unperturbed wave functions ϕ_n, and proceed as before to find the different approximations to the eigenvalues.

A case often presented is that of the calculation of the transition probability from an initial state to a group of very closely spaced states. Therefore, the final states can be expressed by a density of states function $\rho(E)$ (Section 1.6). It is found that the transition rate W in this case is given by [1]:

$$W = \frac{2\pi}{\hbar}\rho(E)|H'_{nk}|^2 \qquad (2.26)$$

Eq. (2.26) is usually called the Fermi rule or *Golden rule* of quantum mechanics because of its usefulness. It states that the rate of transitions between one initial state m to a dense group of final states k is given by $2\pi/\hbar$ times the density of final states function multiplied by the square of the matrix element connecting the initial and final states.

Specially interesting is the case for which the time dependence of the perturbation is of the harmonic type, i.e.

$$H'(t) = H_0' e^{\pm i\omega t} \tag{2.27}$$

where H' could correspond to a perturbation from an electromagnetic field. Then, the transitions from one state to another takes place preferably when a quantum of energy $\hbar\omega$ is either absorbed (resonance absorption) or emitted by the system (stimulated emission).

2.3. FREE ELECTRON MODEL OF A SOLID. DENSITY OF STATES FUNCTION

The simplest model explaining the behaviour of electrons in solids, especially in metals, assumes that the valence electrons of the atoms can move freely inside the material. Therefore, according to this model, electrons would behave as a gas enclosed in a box. This model also assumes that the "free" electrons experience a constant electric potential, except at the surfaces of the solid in which a potential barrier of height H prevents them from escaping. From a quantum-mechanical point of view, the model is similar to that of a particle (a fermion in this case) in a three-dimensional square potential well. This relatively simple quantum model was proposed by Sommerfeld, based on the classical Drude's model.

Since the potential energy is constant inside the box, the time-independent Schrödinger equation for electrons would simply be equal to Eq. (2.10) in which for simplicity we can take $V = 0$. We have seen already in Section 2.2.2 that the wave functions are plane waves of the form:

$$\psi(\vec{r}) = \frac{1}{\sqrt{V}} e^{i\vec{k} \cdot \vec{r}} \tag{2.28}$$

where in three dimensions the amplitude of the wave function has been written as $1/\sqrt{V}$, V being the volume of the potential box. In this way the wave function has been normalized, since the probability of finding the electron inside the volume V is one, that is,

$$\int \psi^* \psi \, dV = A^2 \int \exp(-i\vec{k} \cdot \vec{r}) \exp(i\vec{k} \cdot \vec{r}) dV = 1 \tag{2.29}$$

and therefore $A = 1/\sqrt{V}$.

In order to further proceed with the solution of the problem, we have to establish the boundary conditions. In general, in solid state physics, two kinds of boundary conditions are considered. The *fixed or box boundary conditions*, establish that the electron wave function goes to zero at the box surfaces, since the electron cannot escape from the box (we consider that the height of the energy barriers is infinite). This leads to standing wave functions. On the other hand, the periodic or Born-von Karman boundary conditions lead to travelling wave functions for the electrons, which are more convenient for studying the behaviour of electrons in solids. These conditions only impose that the wave function has the same value at the external boundaries of the solid. For simplicity, let us imagine that the electron is inside a cube of edge L. Therefore, according to the *periodic boundary conditions*:

$$\psi(x, y, z) = \psi(x + L, y, z)$$

$$\psi(x, y, z) = \psi(x, y + L, z) \tag{2.30}$$

$$\psi(x, y, z) = \psi(x, y, z + L)$$

Imposing the periodic boundary conditions to the wave function given by Eq. (2.28) we obtain:

$$e^{ik_x L} = 1, \quad e^{ik_y L} = 1, \quad e^{ik_z L} = 1 \tag{2.31}$$

from which the only allowed values for k_x, k_y, k_z are:

$$k_x = \frac{2\pi}{L} n_x, \quad k_y = \frac{2\pi}{L} n_y, \quad k_z = \frac{2\pi}{L} n_z \tag{2.32}$$

where $n_x, n_y, n_z = 0, \pm 1, \pm 2, \ldots$

Since the wave vectors are quantified according to Eq. (2.32), the corresponding quantification for the energy is:

$$E(n_x, n_y, n_z) = \frac{h^2}{2m}(k_x^2 + k_y^2 + k_z^2) = \frac{h^2}{2m}\left(\frac{2\pi}{L}\right)^2 (n_x^2 + n_y^2 + n_z^2) \tag{2.33}$$

with $n_x, n_y, n_z = 0, \pm 1, \pm 2, \ldots$

In the above discussion, every set of numbers (n_x, n_y, n_z) represents a state of the electron of wave number and energy given by Eqs (2.32) and (2.33), respectively. Although the ideal situation in a solid would be to know all the energies and wave functions of all the states, often it is sufficient to know the *density of states function* $\rho(E)$(DOS), at a given energy, which was defined in Section 1.6. As we see along the different chapters of this text, most of the important properties of a solid, like electrical and transport properties, will be related to the DOS function.

Every electron state is defined by the set of numbers (k_x, k_y, k_z). According to the Pauli exclusion principle there will be two electrons (spin up and spin down) for each occupied state. Since the electron energy is proportional to k^2, the occupied points in k-space (Figure 2.2), expressed by the set of all combinations of values of k_x, k_y, k_z will be located inside a sphere of radius $k = k_{\text{max}}$. On the other hand, the difference between two consecutive values of each k_i component ($i = x, y, z$) is $2\pi/L$. Therefore, each allowed value of $\vec{k}(k_x, k_y, k_z)$ should occupy a volume in k-space given by

$$\left(\frac{2\pi}{L}\right)^3 = \frac{(2\pi)^3}{V} \tag{2.34}$$

where V is the volume of the crystal. Thus, the number of electron states with values lying between k and $k + dk$ (Figure 2.2) should be

$$2\frac{4\pi k^2 dk}{(2\pi)^3/V} = \frac{Vk^2 dk}{\pi^2} \tag{2.35}$$

where the factor 2 takes into account the spin.

Since we know that the $E = E(\vec{k})$ relation is given by Eq. (2.33), we have finally for the DOS function in energy the expression

$$\rho(E) = \frac{4\pi}{\hbar^3}(2m)^{3/2}E^{1/2} \tag{2.36}$$

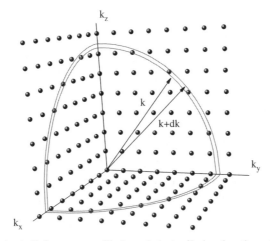

Figure 2.2. Spherical shell between radii k and $k + dk$ in the three-dimensional k-space, corresponding to energies between E and $E + dE$.

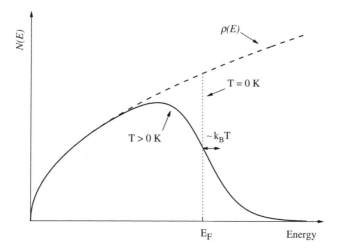

Figure 2.3. Density of occupied electron states as a function of energy.

Observe that, as we have previously mentioned in Section 1.6, $\rho(E)$ increases as the square root of energy. If we multiply $\rho(E)$ by the probability occupation factor given by Eq. (2.17), we can represent the distribution function in energy for the electrons (Figure 2.3). It is evident that at $0\,K$ we only have occupied electron states up to E_F. For $T > 0\,K$, some of the electrons with energies below E_F are promoted to states above E_F.

2.4. BLOCH THEOREM

Bloch's theorem (1928) applies to wave functions of electrons inside a crystal and rests in the fact that the Coulomb potential in a crystalline solid is periodic. As a consequence, the potential energy function, $V(\vec{r})$, in Schrödinger's equation should be of the form:

$$V(\vec{r}) = V(\vec{r} + \vec{R}_n)$$

(2.37)

where $\vec{R}_n$ represents an arbitrary translation vector of the crystallographic lattice, i.e. $\vec{R}_n = n_1\vec{a}_1 + n_2\vec{a}_2 + n_3\vec{a}_3$, ($\vec{a}_1, \vec{a}_2, \vec{a}_3$ are the unit lattice vectors).

Bloch's theorem establishes that the wave function $\psi_{\vec{k}}(\vec{r})$ in a crystal, obtained from Schrödinger's Eq. (2.10), can be expressed as the product of a plane wave and a function $u_{\vec{k}}(\vec{r})$ which has the same periodicity as the lattice, i.e.

$$\psi_{\vec{k}}(\vec{r}) = e^{i\vec{k}\cdot\vec{r}} u_{\vec{k}}(\vec{r})$$

(2.38)

where

$$u_{\vec{k}}(\vec{r}) = u_{\vec{k}}(\vec{r} + \vec{R}_n) \tag{2.39}$$

The electron wave functions, of the form of Eq. (2.38), are called *Bloch functions*. Note that although the Bloch functions are not themselves periodic, because of the plane wave component in Eq. (2.38), the probability density function $|\psi_{\vec{k}}|^2$ has the periodicity of the lattice, as it can be easily shown. Another interesting property of the wave functions derived from Bloch's theorem is the following:

$$\psi_{\vec{k}}(\vec{r} + \vec{R}_n) = e^{i\vec{k}\cdot\vec{R}_n}\psi_{\vec{k}}(\vec{r}) \tag{2.40}$$

It can be appreciated that this property is a direct consequence of Eqs (2.38) and (2.39).

2.5. ELECTRONS IN CRYSTALLINE SOLIDS

2.5.1. *Nearly free electron model*

The free electron model in solids discussed in the previous section is very useful in order to explain some of the electrical properties of metals, but it is in general too simple to explain many of the electronic and optical properties of semiconductors. A normal extension of this model is the *nearly free electron model (NFE)*, which applies to crystals, since it is assumed that the electrons are subjected to a periodic potential inside them. Figure 2.4 shows the representation, in one dimension, of the potential inside a crystal. Figure 2.4(a) represents the potential along a line of nuclei in the 3D crystal, the Coulombic potentials evidently approaching $-\infty$ at the nuclei positions. However, most of the electrons in the 3D crystal move along lines located between crystallographic directions going through the nuclei. For instance, for a direction located midway between two planes of ions, the potential should be similar to the one shown in Figure 2.4(b).

As a consequence of the periodicity of the crystal, we will see that the electrons can only have values of energy in certain allowed regions or bands, while some other energy intervals will be forbidden. In the NFE model, the periodic potential is considered as a small perturbation to the Hamiltonian corresponding to the free electron model. Although the approximation is a little rough, the qualitative consequences of this simple model are very important. In this text we will not try to fully develop the NFE model, and therefore we will address readers to standard solid state physics books [2]. However, we will present its main consequences. One of them, due to the periodicity of the crystal, is that the electron wave functions ψ are of the Bloch type (Section 2.4), that is:

$$\psi_k = u_k(x)e^{ikx} \tag{2.41}$$

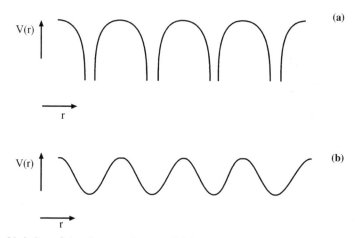

Figure 2.4. Variation of the electrostatic potential for a simple cubic lattice. (a) Potential through a line of atoms; (b) idem through a direction between lines of atoms.

where u_k should have the periodicity of the lattice. (For simplicity we assume a 1D model of lattice constant a.)

Since the periodic potential inside the crystal is taken as a perturbation, it is necessary to use quantum mechanical perturbation theory (Section 2.2.4). The solution of this problem by perturbative techniques gives us the result that there will be discontinuities in the energy values whenever

$$k = \frac{\pi}{a}n, \qquad n = \pm 1, \pm 2, \dots \tag{2.42}$$

The values of the discontinuities in energy, or energy gaps, are proportional to the coefficients $|V_n|$ of the Fourier expansion of the potential [2], i.e.

$$V(x) = \sum_{n=-\infty}^{+\infty} V_n e^{i\frac{2\pi}{a}nx} \tag{2.43}$$

The *energy bands and gaps* for the above 1D model are shown in Figure 2.5. This is the usual *extended zone representation*, in which it can be appreciated the similarity of the $E = E(k)$ curve with the free electron parabola, except at the values of k given by Eq. (2.42), where gaps open up. Figure 2.6(a) is the so-called *reduced zone representation*, which in reality is a consequence of the periodicity of the wave functions in reciprocal space. This representation is the most often used and can be obtained from that

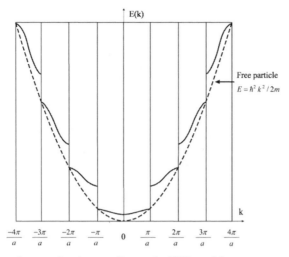

Figure 2.5. Structure of energy bands according to the NFE model.

of Figure 2.5 by translations of multiples of $2\pi/a$ along the k direction. Figure 2.6(b) is called the *repeated zone representation* and is the least used since the information it gives is redundant.

2.5.2. Tight binding approximation

The nearly free electron model of Section 2.5.1 is not appropriate for the description of insulating materials like, for instance, diamond. In these cases, the *tight binding (TB) approximation* yields better results. This approximation starts by considering, contrary to the NFE model, that the potential energy of electrons in the atom is the main component of the total energy and it is assumed that the wave functions of the electrons in two neighbouring atoms have little overlap. The TB approximation, first proposed by Bloch, is similar to the linear combination of atomic orbitals (LCAO) method for molecules and works fairly well for the case of the electrons in insulators or for inner shell electrons in metals. The method starts by assuming that the electron wave functions are known for the orbitals of the individual atoms $\phi_0(\vec{r} - \vec{R}_n)$, where $\vec{R}_n$ is a general translation vector of the lattice. Next it is assumed that the wave function of the electron in the crystal is a linear combination of atomic orbitals, i.e.

$$\phi_{\vec{k}}(\vec{r}) = \sum_n C_n \phi_0(\vec{r} - \vec{R}_n) \tag{2.44}$$

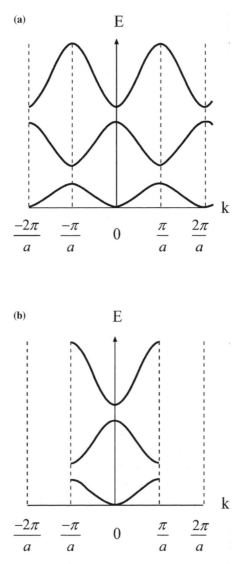

Figure 2.6. (a) Energy as a function of k in the repeated zone representation; (b) idem in the reduced zone representation.

where the sum is extended to the lattice points. This wave function should verify Bloch theorem and therefore the coefficients C_n are of the form $e^{i\vec{k}\cdot\vec{R}_n}$, thus having:

$$\phi_{\vec{k}}(\vec{r}) = \sqrt{N} \sum_{n} e^{i\vec{k}\cdot\vec{R}_n} \phi_0(\vec{r} - \vec{R}_n) \tag{2.45}$$

where N is the number of atoms in the lattice and $N^{1/2}$ is the proper normalizing factor as it can be easily proven.

The TB approximation assumes that the potential experimented by the electron in the solid $V(\vec{r})$ is very close to $V_0(\vec{r})$ in the isolated atom, plus a small perturbation $V'(\vec{r})$. Therefore, the Hamiltonian for the electron in the crystal can be written as

$$H = -\frac{\hbar^2}{2m_e}\nabla^2 + V_0(\vec{r} - \vec{R}) + V'(\vec{r} - \vec{R}) \tag{2.46}$$

Solving the perturbation problem with the Hamiltonian of Eq. (2.46) and the wave functions of Eq. (2.45), it can be shown [3] that:

$$E(\vec{k}) = E_0 - \alpha - \beta \sum_m e^{i\vec{k}\cdot(\vec{R}_m - \vec{R}_n)} \tag{2.47}$$

where the sum is over the nearest neighbours. In Eq. (2.47), α and β are the so-called overlap integrals, which are often treated as adjustable parameters.

As an example, let us apply Eq. (2.47) to the case of a simple cubic lattice of parameter a considering the six-nearest neighbours:

$$\vec{R}_m - \vec{R}_n = (\pm a, 0, 0), (0, \pm a, 0), (0, 0, \pm a) \tag{2.48}$$

then:

$$\vec{E}(\vec{k}) = E_0 - \alpha - 2\beta(\cos k_x a + \cos k_y a + \cos k_z a) \tag{2.49}$$

Observe that the width of the energy band is equal to 12β, i.e. the width depends on the overlapping integral, which takes into account the overlap between neighbouring wave functions and decreases rapidly with the distance between atoms. The minimum energy occurs at $k = 0$, and the maximum at the boundary ($k_x = \pi/a$, $k_y = \pi/a$, $k_z = \pi/a$) of the Brillouin zone. Around the point $\vec{k} = (k_x, k_y, k_z) = 0$, Eq. (2.49) can be expanded in series obtaining:

$$E(k) = E_0 - \alpha - 6\beta + \beta k^2 a^2 \tag{2.50}$$

Note that near $k = 0$, the dependence of energy on k is quadratic and the constant energy surfaces are spherical. From Eq. (2.50) we can calculate the electron effective mass (Section 2.6.2), yielding a value $m^* = \hbar^2/2\beta a^2$. Observe that m^* increases when the overlap integral β decreases. In the limit of isolated atoms m^* is infinite as expected, since an external force provided by an electric field cannot translate the electron from one lattice site to another.

2.6. DYNAMICS OF ELECTRONS IN BANDS

2.6.1. *Equation of motion*

We have seen that the wave function of a free electron is a plane travelling wave. This wave function represents a particle of a well-defined momentum $p = \hbar k$. However, as a consequence of Heisenberg's uncertainty principle, this wave given by Eq. (2.14) cannot tell us anything about the localization of the electron in space. For this reason, if we want to describe the position and momentum of an electron inside a crystal, we have to make use of *wave packets*. The usual way to form wave packets is by means of the lineal superposition of plane travelling waves with wave vectors included in a small interval Δk around a mean value k, i.e. the wave function should be of the form:

$$\psi(x, t) \propto \int_{k-\Delta k/2}^{k+\Delta k/2} c(k)\, e^{i[kx-\omega t]} dk \tag{2.51}$$

where, in general, $\omega = \omega(k)$, for dispersive media. The wave packet moves with the group velocity v_{g} given by:

$$v_{\mathrm{g}} = \frac{\partial \omega}{\partial k} \tag{2.52}$$

which in general differs from the phase velocity ($v = \omega/k$) of the plane waves.

Evidently, if the electron moves inside a crystal and we adopt the nearly free electron model (Section 2.5.1) for the solid, the wave function of the electron can be expressed in a three-dimensional crystal as a Fourier series

$$\psi = \sum_{k} c(k, t) e^{i(\vec{k}\cdot\vec{r} - \omega t)} \tag{2.53}$$

where the values of $\vec{k}$ (Section 2.3) are defined by the periodic boundary conditions. Extending Eq. (2.52) to three dimensions, the group velocity is given by:

$$\vec{v}_{\mathrm{g}} = \vec{\nabla}_{\vec{k}}\, \omega\left(\vec{k}\right) = \frac{1}{\hbar}\vec{\nabla}_{\vec{k}} E\left(\vec{k}\right) \tag{2.54}$$

where $E = E(\vec{k})$ is the relation between energy and momentum for an electron in a given band.

Suppose now that we apply a force to an electron inside the crystal by means of an electric field $\vec{F}$ and let us adopt a one-dimensional model for simplicity. The work δE performed on the electron of charge $-e$ by the field $\vec{F}$ during the time integral δt is, in one dimension

$$\delta E = -eFv_g\delta t \tag{2.55}$$

but, from Eq. (2.54), the above expression can be written as:

$$\delta E = \left(\frac{dE}{dk}\right)\delta k = \hbar v_g\, \delta k \tag{2.56}$$

Generalizing to three dimensions, we have the following expression for the force on the electron:

$$-e\vec{F} = h\frac{d\vec{k}}{dt} \tag{2.57}$$

Eq. (2.57) is the *equation of motion* of the electron. Observe that it looks similar to Newton's second law for free electrons, if we assign a momentum $\vec{p}$ to the electron given by

$$\vec{p} = \hbar\vec{k} \tag{2.58}$$

The momentum given by Eq. (2.58) is not the true momentum since in reality the electron not only interacts with the electric field, as we have considered in the derivation of Eq. (2.57), but is also subjected to the Coulombic forces of the lattice ions. In fact, inside the crystal, is impossible to treat individually each force acting on the electron, as if it were isolated. Therefore, the momentum given by Eq. (2.58) is called the *crystal momentum* and its full meaning is evidenced when the electrons interact with other particles such as electrons or phonons (Section 2.8), the crystal momentum being conserved, together with energy, in these interactions.

2.6.2. *Effective mass*

In order to find for an electron in a crystal, the expression corresponding to the concept of mass of a particle, let us find first the relation between force and acceleration. In effect, let us apply to an electron in a crystal a force caused by an electric field F as we did previously. According to Eqs (2.52) and (2.57), the expression for the acceleration should be, in one dimension,

$$\frac{dv}{dt} = \frac{1}{\hbar}\frac{d^2 E}{dt\, dk} = \frac{1}{\hbar}\frac{d^2 E}{dk^2}\frac{dk}{dt} = \frac{1}{\hbar^2}\frac{d^2 E}{dk^2}(-eF) \tag{2.59}$$

From this relation, we define the *effective mass* m^* of the electron in the crystal by the expression:

$$\frac{1}{m^*} \equiv \frac{1}{\hbar^2}\frac{d^2 E}{dk^2} \tag{2.60}$$

Note that m^* is inversely proportional to the curvature of the energy band, i.e. the flatter the band in k-space, the larger the value of the effective mass.

In a three-dimensional crystal, the *effective mass tensor* has components

$$\left(\frac{1}{m^*}\right)_{ij} \equiv \frac{1}{\hbar^2}\frac{\partial^2 E}{\partial k_i \partial k_j} \tag{2.61}$$

Observe that in a crystal the force and acceleration do not have, in general, the same direction. In the simplest case, the three effective masses, corresponding to the principal axes of the tensor of Eq. (2.61) are equal, and the expression for m^* is similar to Eq. (2.60) for the one-dimensional crystal, i.e. m^* becomes a scalar. This should happen when the E vs k relation has a parabolic dependence along every axis in k-space, i.e.

$$E(\vec{k}) = E_0 + \frac{\hbar^2}{2m^*}(k_x^2 + k_y^2 + k_z^2) \tag{2.62}$$

From Eq. (2.62) it can be observed that in this case the surfaces of constant energy, $E = E(\vec{k})$ are spheres.

According to the definition of effective mass, we can note that if the dependence of E on the wave vector has the shape shown in Figure 2.7(a), then m^* from Eq. (2.60) is represented in Figure 2.7(b). Observe that m^* is positive at the bottom of the band but negative (curvature of the opposite sign) for k values close to the zone boundaries $k = \pm\pi/a$. This means that if the electron is for instance approaching the boundary $k = \pi/a$ from the left, it cannot gain any more momentum from the electric field, or better, the electron transfers more momentum to the lattice than what it gains from the applied field. In the limit, at the zone boundaries ($k = \pm\pi/a$), the electron gets Bragg reflected.

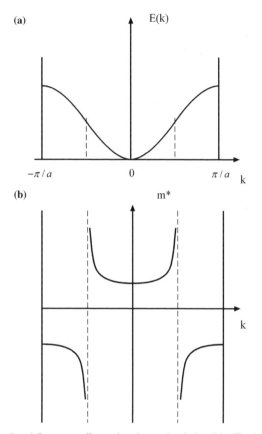

Figure 2.7. (a) Energy band for a one-dimensional atomic chain; (b) effective mass deduced from the above energy band.

2.6.3. *Holes*

In the study of the behaviour of electrons in a band which is nearly filled, as it frequently happens in semiconductors, it is convenient to introduce the concept of hole. For simplicity, let us consider first the case of a band like the lower one in Figure 2.8 whose states are all full, except for one electron with wave vector $\vec{k}_e$. The total wave vector $\vec{k}_T$ of this band can be written as the wave vector of a completely filled band, minus the wave vector of the missing electron, i.e.

$$\vec{k}_T = \sum_{k_i} \vec{k}_i - \left(\vec{k}_e\right) \tag{2.63}$$

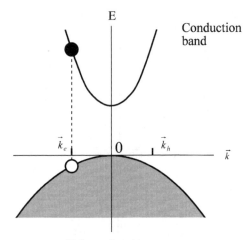

Figure 2.8. Hole in the valence band of a direct gap semiconductor.

The sum in Eq. (2.63) is extended to the full band and should be zero if the crystal has inversion symmetry. In this case, the band in k-space has also inversion symmetry, and for each occupied state $\vec{k}_i$ there is also one with wave vector $-\vec{k}_i$. Therefore from Eq. (2.63) we have

$$\vec{k}_T = -\vec{k}_e \qquad (2.64)$$

The state of the missing electron can be assigned in the band to a new particle, known as *hole*, with a wave vector

$$\vec{k}_h = -\vec{k}_e \qquad (2.65)$$

and momentum

$$\vec{p}_h = -\hbar\vec{k}_e = \hbar\vec{k}_h \qquad (2.66)$$

Since the concept of hole appears when the band is full except for one electron, it should have a positive charge. In a semiconductor holes are often created if one photon transfers its energy to an electron raising it from a state in the lower band to a state in the upper band as shown in Figure 2.8. If the energy of the photon increases, holes with higher energies can be created. Therefore the energy of a hole is higher as we move down the band. The kinetic energy E_h of a hole is positive and from the relation $E_h = \hbar^2 k^2 / 2m_h^*$, it is appreciated that the effective mass of a hole should be also positive. In summary, note that the concept of hole has been conveniently introduced, since it is easier to describe an almost filled band by a few empty states, instead of by means of many occupied states.

2.7. LATTICE VIBRATIONS

In this section we pretend to give a short review of vibrations in periodic systems such as crystals. The "adiabatic approximation" in solid state physics allows the separate study of those properties of materials, attributed to electrons, like the electrical conductivity, and those which depend on the vibrations of the atoms, such as the thermal properties. Suppose a mechanical wave, or a sound wave, travelling through a solid. If its wavelength λ is much larger than the lattice constant of the crystal, then the medium behaves as an elastic continuum, although not necessarily isotropic. However, when λ is comparable or smaller than the lattice unit cell, we have to consider the crystalline structure of the solid.

In order to treat the vibrations of the lattice atoms, we will usually follow the harmonic approximation. Forces between neighbouring atoms have their origin in a kind of potential which is mainly attractive, giving rise to the interatomic bonding (e.g. covalent, ionic, van der Waals). However, if the distance between atoms becomes very small, the electrons between two neighbouring atoms start to interact, and because of the Pauli exclusion principle, a repulsive interaction appears which increases very rapidly as the distance between them decreases. One of the better-known potentials which describes this interaction is the Lennard–Jones potential. This potential shows a minimum when the interatomic distance r is equal to the one at equilibrium, i.e. to the lattice constant a. For values of r close to a, the potential $V(r)$ can be approximated by a parabolic or harmonic potential. At the beginning, we will assume that this *harmonic approximation* is valid around $r = a$.

2.7.1. One-dimensional lattice

The simplest model to study vibrations in a periodic solid is known as the *one-dimensional monoatomic chain*, which consists of a chain of atoms of mass m, equilibrium distance a, and harmonic interaction between atoms (Figure 2.9(a)). In this figure we call u_n the displacement of the atoms from the equilibrium position. The equation of motion of atom n, if we only consider interaction between closest neighbours, should be:

$$m \frac{d^2 u_n}{dx^2} = c\left[(u_{n+1} - u_n) - (u_n - u_{n-1})\right] = c(u_{n+1} + u_{n-1} - 2u_n) \tag{2.67}$$

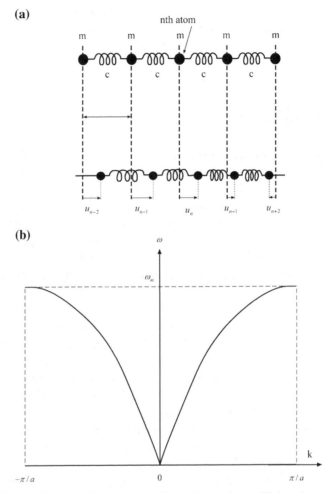

Figure 2.9. (a) One-dimensional monoatomic chain of atoms in equilibrium (upper) and displaced from equilibrium (lower); (b) representation of the dispersion relation.

and similarly for any other atom in the chain. In order to solve Eq. (2.67), travelling plane waves of amplitude A, frequency ω and wave number k are assumed, i.e.

$$u_n = Ae^{i(kx_n - \omega t)} \tag{2.68}$$

where $x_n = na$ is the equilibrium position of the atoms. Substituting Eq. (2.68) for the atomic displacement u_n and the corresponding ones for u_{n+1} and u_{n-1}, after some algebra, one gets [2] from Eq. (2.67).

$$\omega = \left(\frac{4c}{m}\right)^{1/2} \left|\sin\frac{ka}{2}\right| \qquad (2.69)$$

Eq. (2.69), known as the *dispersion relation* for vibrations in a one-dimensional lattice, is represented in Figure 2.9(b). One important consequence of this equation, difficult to imagine in continuous media, is the existence of a maximum frequency of value $2(c/m)^{1/2}$, over which waves cannot propagate. This frequency is obtained when $k = \pi/a$, i.e. $\lambda = 2a$ in Eq. (2.69). This condition is similar to that for a Bragg reflection for electrons in periodic structures (Section 2.6.2), and mathematically leads to standing waves, instead of travelling waves, which cannot propagate energy. This result should be expected since for $k = \pi/a$, the group velocity is zero. Note also in Eq. (2.69) that for $ka \to 0$, ω varies linearly with k and the group velocity coincides with the phase velocity of the wave, both having the value $v_s = a(c/m)^{1/2}$. This result is expected since for $ka \to 0$, $a/\lambda \to 0$, i.e. the wavelength is much greater than the interatomic distance, and the medium can be considered as continuous. In this situation v_s is equivalent to the speed of sound in the medium.

Following a similar procedure as in the case of electrons in periodic crystals (Section 2.3), we can establish periodic boundary conditions for the solutions given by the waves of Eq. (2.68). Physically these conditions could be obtained by establishing a fixed link or constraint forcing the first and last atoms of the chain to perform the same movement. This results (see Eq. (2.32)) in the following allowed values for k:

$$k = \frac{2\pi}{L} n, \quad n = 0, \pm 1, \pm 2, \ldots \pm N \qquad (2.70)$$

where L and N are the length of the chain and the total number of atoms, respectively, i.e. $L = Na$.

A final important consequence of the dispersion relation is that the value of ω remains the same whenever the value of k changes in multiples of $2\pi/a$. Therefore, it would be sufficient, as it happens for electrons, to consider only the values of k belonging to the first Brillouin zone, that is

$$-\frac{\pi}{a} \leq k \leq \frac{\pi}{a} \qquad (2.71)$$

The next level of complexity in the study of lattice vibrations comes when the crystal has more than one atom per primitive unit cell. Suppose then the *diatomic linear chain* of Figure 2.10(a) with two kinds of atoms of masses M and m, where $M > m$. The main difference arises now from the fact that the amplitudes of the atoms M and m are unequal.

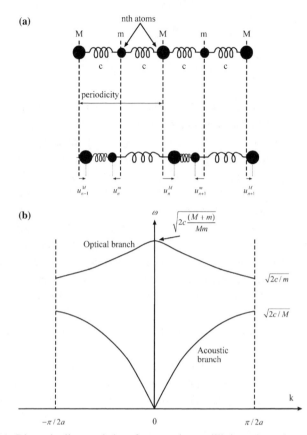

Figure 2.10. (a) Diatomic linear chain of atoms in equilibrium (upper) and displaced from equilibrium (lower); (b) representation of the dispersion relation.

If one amplitude is A, the other is αA, where α is in general a complex number that takes into account the relation between amplitudes as well as the phase difference. We can next proceed in an analogous way as we did with the linear chain. The solution of the problem becomes now more complicated but the calculation is straightforward and can be found in any elementary text on solid state physics.

For the diatomic linear chain, the *dispersion relation* is found to be [2]:

$$\omega^2 = \frac{c(m+M)}{Mm} \pm k\left[\left(\frac{M+m}{Mm}\right)^2 - \frac{4}{Mm}\sin^2\left(\frac{ka}{2}\right)\right]^{1/2} \qquad (2.72)$$

The dispersion relation, represented in Figure 2.10(b), has now two branches, the upper and the lower corresponding to the $\pm$ signs of Eq. (2.72), respectively. As for the monoatomic chain, it is instructive to examine the solutions at certain values of k, close to the centre zone ($k = 0$) and at the boundaries ($k = \pi/a$). If $ka \ll 1$, then $\alpha = 1$ for the lower or *acoustic branch* and $\alpha = -M/m$ for the upper or *optical branch*. If $\alpha = 1$, the neighbouring atoms M and m vibrate essentially with the same phase, as it happens with the sound waves in solids for which $\lambda \gg a$. If $\alpha = -M/m$, both particles oscillate out of phase, and the upper branch presents a maximum frequency equal to:

$$\omega_{max}^+ = \left(\frac{2c\,(M + m)}{Mm} \right)^{1/2} \tag{2.73}$$

The vibrational modes of this branch are called optical, because the value of the frequency is in the infrared range and in crystals such as NaCl, with a strong ionic character, the optical modes can be excited by an electromagnetic radiation. In these modes, the positive and negative ions move evidently out of phase when excited by the oscillating electric field of an electromagnetic radiation.

2.7.2. Three-dimensional lattice

The dispersion relations considered in Section 2.7.1 for the case of the 1D crystal can now be generalized to three dimensions [4]. The number of acoustic branches for a 3D lattice is three, one longitudinal in which the atoms vibrate in one direction of the chain and two transverse ones, in which the atoms vibrate perpendicularly to the direction of the propagation of the wave. Therefore, there will be one longitudinal acoustic branch (LA) and two transverse acoustic branches (TA) which are often degenerated. As a note of caution, we would like to remark that one has to be careful with the above statement regarding the directions of the vibrations of the atoms, because if $\vec{k}$ is not along a direction with high symmetry, then the atomic displacements are not exactly along $\vec{k}$ or perpendicular to it.

As in 1D, if there is more than one atom, let us say p per primitive unit cell, the number of optical branches is, in general, $3p-3$. If $p = 2$, as for an example in alkali halides, there are three acoustic branches and three optical ones. However in highly symmetric directions the two transverse modes might be degenerated. Figure 2.11 shows schematically the dispersion relations for a 3D crystal. Since there are no degeneracies we have assumed that the crystal is anisotropic. Note also that in general, as it happens with electrons in crystals, the dispersion curves cut the Brillouin zone boundaries perpendicularly, although there might be exceptions in the case of very complicated shapes of the Brillouin zones.

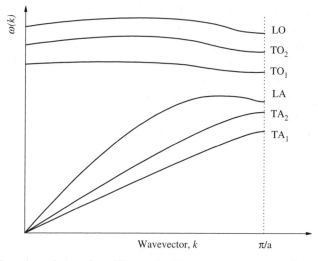

Figure 2.11. Dispersion relations for a 3D crystal with two atoms per unit cell.

2.8. PHONONS

In the study of the lattice vibrations we have assumed so far that the interaction between neighbouring atoms was harmonic, and under this consideration we wrote the equations of motion of the atoms like Eq. (2.67). This equation is somewhat similar to the equation of an individual harmonic oscillator for u_n:

$$m\frac{d^2u_n}{dt^2} = -cu_n \qquad (2.74)$$

except that in the second term of Eq. (2.67) there also enter terms corresponding to neighbouring atoms $n-1$ and $n+1$. It is known from classical mechanics that in this case it is possible to find the *normal coordinates* of this system of particles, i.e. appropriate linear combinations of u_n such that the equations of motion have the simple form of Eq. (2.74). The advantage of using normal coordinates is that the Hamiltonian of the system becomes diagonal, i.e. we can write the Hamiltonian as a sum of individual or non-coupled Hamiltonians for harmonic oscillators.

After having performed the transformation to normal coordinates, it is easier to consider the problem from a quantum-mechanical point of view. In effect, as a consequence of the decomposition of the Hamiltonian of the system as a sum of uncoupled individual Hamiltonians, the quantum-mechanical states can be expressed as the product of wave functions of harmonic oscillators, one for each normal mode. Each *normal mode* of frequency ω_k has the following possible values of energy E_k given by

$$E_k = (n_k + \frac{1}{2})\hbar\omega_k, \quad n_k = 0, 1, 2, \ldots \tag{2.75}$$

This means, that the state k has n_k quanta of energy $\hbar\omega_k$. These quanta of energy that arise from the lattice vibrations are called *phonons*. If the normal mode k is excited from n_k to $n_k + 1$ we say that one phonon of energy $\hbar\omega_k$ has been gained by the system and similarly for the loss of phonons. The creation and annihilation of phonons is possible because phonons are quasi-particles and their total number does not have to be conserved. Taking into account the above considerations, the total energy E of the lattice vibrations of the system can be expressed as

$$E = \sum_{k,p} \hbar\omega_{k,p}(n_{k,p} + \frac{1}{2}) \tag{2.76}$$

where the summation is extended to all acoustic and optical branches.

REFERENCES

[1] Merzbacher, E. (1960) *Quantum Mechanics* (Wiley, New York).
[2] Kittel, Ch. (2005) *Introduction to Solid State Physics*, 8th edition (Wiley).
[3] Ibach, H. & Luth, H. (1991) *Solid-State Physics* (Springer-Verlag, Berlin).
[4] Elliot, S. (1998) *The Physics and Chemistry of Solids* (Wiley, Chichester).

FURTHER READING

Some excellent books on solid state physics, at the intermediate level, are the following:
Burns, G. (1995) *Solid State Physics* (Academic Press, Boston).
Elliot, S. (1998) *The Physics and Chemistry of Solids* (Wiley, Chichester).
Ibach, H. & Luth, H. (1991) *Solid-State Physics* (Springer-Verlag, Berlin).
Kittel, Ch. (2005) *Introduction to Solid State Physics*, 8th edition (Wiley).
Singleton, J. (2001) *Band Theory and Electronic Properties of Solids* (Oxford University Press, Oxford).

PROBLEMS

1. **Fermi energy.** Suppose that the behaviour of metallic sodium can be explained by the free electron model. (a) Calculate the Fermi energy (the density of sodium is of 0.97 g cm^{-3}). (b) Calculate the average energy per electron.

2. **Electronic bands in solids**. Show that the energy bands of a crystal, derived either from the nearly-free electron model or the tight-binding approximation, have many points in common. In particular, show: (a) The bands have minima (electron states) and maxima (hole states). (b) Around the maxima and minima the bands are parabolic in k. (c) The bands are periodic in reciprocal space, and therefore all the information is included in the first Brillouin zone.

3. **Effective mass in a 1D solid**. In a particular 1D lattice, the electron band energy can be written as

$$E(k) = \frac{\hbar^2 k^2}{2m_0^*} - \alpha k^4$$

Determine the value of the constant α, the effective mass for $k = 0$ and $k = \pi/a$, the maximum velocity and the corresponding energy, and the effective mass at the centre and the edge of the Brillouin zone. *Hint*: first determine α considering that $v_g = 0$ at π/a.

4. **Effective mass tensor**. Considering that the electronic energy band of a simple cubic lattice of side a within the tight-binding approximation is given by

$$E = -E_1(\cos\, k_x a + \cos\, k_y a + \cos\, k_z a),$$

calculate the effective mass tensor m_{ij} at the centre in k-space, at the face centre and at the corner. *Hint*: realize first that the three points in which you have to determine the effective mass have coordinates in k-space given by $k = \pi/a(0, 0, 0)$, $k = \pi/a(0, 0, 1)$, and $k = \pi/a(1, 1, 1)$.

5. **Bragg condition in 2D**. Consider a square lattice of identical atoms and side a. Show that a plane wave verifies Bragg reflection at the first Brillouin zone boundaries.

6. **Dispersion curves in lattice vibrations**. Considering a unidimensional lattice with two molecules in each primitive cell, calculate the characteristic frequencies. Consider a polyethylene chain (–CH=CH–CH=CH–) of identical masses of value M, connected alternatively by springs of constants C_1 and C_2, respectively and only interaction to first neighbours. Plot the resulting dispersion curves for the acoustic and optical branches. *Hint*: first show that the characteristic frequencies are given by:

$$\omega^2 = \frac{C_1 + C_2}{M}\left[1 \pm \left(1 - \frac{4C_1 C_2 \sin^2(ka/2)}{(C_1 + C_2)^2}\right)^{1/2}\right]$$

7. **Electron wave vector**. Ideally assuming 100% efficiency, estimate the change in wave vector associated to the dispersion of an electron from the valence to the conduction band as a consequence of the absorption of a photon, and compare the resulting values with the typical first Brillouin zone dimensions, in the following cases: (a) a blue CdSe laser ($E_g = 2.7\,\text{eV}$), GaAs ($E_g = 1.6\,\text{eV}$), and (b) a HgCdTe compound ($E_g = 0.15\,\text{eV}$).

Chapter 3

Review of Semiconductor Physics

Chapter 3

Review of Semiconductor Physics

3.1. INTRODUCTION

In this chapter, we make a review of the physics of bulk semiconductors which we need as previous knowledge for the understanding of the behaviour of semiconductors in mesoscopic systems of low dimensionality (Section 1.5). We will focus mainly on those electronic and optical properties related to the understanding of device applications such as transistors, lasers, etc. The chapter starts with a description of the band structure of typical semiconductors. Next, the calculation of electron and hole concentrations in intrinsic and extrinsic semiconductors is considered. The mechanism of electron transport in semiconductors, both under the action of electric fields and carrier concentration gradients, is also revised. Generation and recombination of carriers in electric fields and under concentration gradients lead to the continuity equation and to the concepts of minority carrier lifetime and diffusion length. The last sections of the chapter are devoted to the study of optical processes in semiconductors, especially to the processes of light absorption, light emission, and exciton transitions, since they constitute the basis of many optoelectronic devices like lasers, modulators, etc.

3.2. ENERGY BANDS IN TYPICAL SEMICONDUCTORS

The parameter which determines most of the properties (electrical, optoelectronic, etc.) of semiconductors is the energy gap, or forbidden energy region between the last completely filled band, known as the valence band, and the next higher energy band, or conduction band, which can be empty or partially filled. In addition to the gap, it is important to know the curvature of these bands in k-space, which determines the effective mass of the carriers (electrons or holes). As we have seen in the previous chapter, in many solids, the shape of energy bands around their maxima and minima in k-space can be considered parabolic in a first approximation, a fact which simplifies the description of the semiconductors by the effective mass approximation. The direct or indirect character of the gap has also a strong influence in many of the properties of semiconductors, like for instance, those related to optoelectronic applications.

   Most of the important semiconductors from a technological point of view, like III-V compounds, Si, Ge, etc., have cubic symmetry. In these semiconductors the maximum of the valence band is located at $\vec{k} = 0$. However, the minimum of the conduction band is either located at $\vec{k} = 0$, as in GaAs, or can be situated close to the border of the first

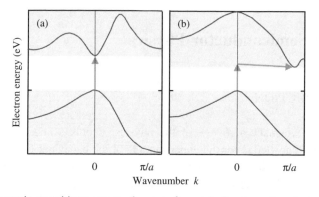

Figure 3.1. Electronic transitions across the gap for: (a) direct semiconductor; (b) indirect semiconductor.

Brillouin zone, as in the case of Si or Ge. Semiconductors of the first type are said to have a *direct gap*, since the absorption or emission of photons involves the transfer of one electron from one band to the other without changing the wave vector, i.e. the transition can be represented by a vertical line in k-space like the one shown in Figure 3.1(a). In this case, the wave number (and therefore the momentum) of the electron after the transition is practically the same as the initial one, since the wave number k of the photons ($k = p/\hbar$) is practically negligible in comparison with those of the electrons. Suppose now an electron transition between the maximum of the valence band to the minimum of the conduction band in an *indirect gap* semiconductor like Si or Ge (Figure 3.1(b)). In this transition, the electron has to change its wave vector by a large amount, almost π/a. Therefore for electrons to absorb or emit photons, its momentum has also to be changed by a large amount. This needs the participation of a third particle which emits or absorbs the difference in momentum. Evidently this particle is a phonon (Section 2.8) of the appropriate value of momentum. Since the participation of a third particle is needed, the probability of photon emission is much lower for indirect gap semiconductors in comparison to direct ones. For this reason, optoelectronic devices such as light emission diodes or semiconductors lasers are built from direct gap semiconductors.

Figures 3.2(a) and (b) show the energy bands of gallium arsenide and silicon, respectively. It can be seen that GaAs is direct and has a gap of 1.43 eV whereas silicon is an indirect gap semiconductor with a gap of value $E_G = 1.1$ eV. Since the relation $E = E(\vec{k})$ cannot be visualized in three dimensions, it is usually represented for some high symmetry directions in k-space. For instance, in Figure 3.2 the right axis corresponds to the (1,0,0) direction while the left axis corresponds to the (1,1,1) direction.

In order to calculate the effective masses, we have to make use of Eq. (2.61) of Chapter 2 for the effective mass tensor. In this way we can calculate the effective mass

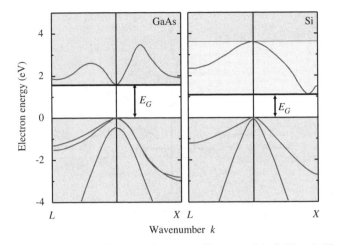

Figure 3.2. Energy bands of typical semiconductors: gallium arsenide (left) and silicon (right).

for motion along any pre-determined direction. In the case of GaAs at point Γ, the $E = E(\vec{k})$ dependence is the same along the three main directions, and therefore the following approximation is valid:

$$E(k) = \frac{\hbar^2 k^2}{2m_e^*} \tag{3.1}$$

which indicates that the tensor reduces to a scalar. We can appreciate in Figure 3.2 the large curvature of the energy band at Γ which implies a small value of the effective mass $(0.066m_0)$. As also seen from Eq. (3.1) the surfaces of constant energy, i.e. $E(\vec{k}) = \text{const}$, are spherical for GaAs.

The equivalent $E = E(\vec{k})$ expression for silicon is a little more complicated, but by choosing the axes properly it can be written as:

$$E(k) = \frac{\hbar^2}{2} \left(\frac{k_l^2}{m_l^*} + \frac{k_t^2}{m_t^*} \right) \tag{3.2}$$

where k_l and k_t are longitudinal and transversal components of $\vec{k}$ and m_l^* and m_t^* are the longitudinal and transversal electron effective masses, respectively. This is due to the fact that the surfaces of constant energy are from Eq. (3.2) revolution ellipsoids as shown in Figure 3.3. As seen from this figure, the six directions $\pm k_x$, $\pm k_y$, $\pm k_z$ are equivalent. The values of the electron effective masses for silicon are $m_l^* = 0.98m_0$ and $m_t^* = 0.066m_0$.

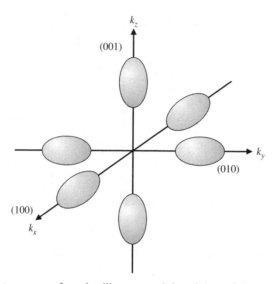

Figure 3.3. Constant energy surfaces in silicon around the minima of the conduction band.

3.3. INTRINSIC AND EXTRINSIC SEMICONDUCTORS

Semiconductors have values of electrical conductivity between those of metals and insulators. Another characteristic of semiconductors is that the electrical conductivity is strongly
dependent on temperature and the level of impurities or dopants. Typical semiconductors
can be elemental (Si, Ge), combination of group III and V elements (GaAs, GaP), II and
VI (ZnS, CdTe), etc. The semiconductor gap usually ranges from a few tenths of eV to
about 3 eV. If the gap is between 2 and 3 eV, the semiconductors are said to have a wide
bandgap, and if it is about 4 eV or larger, the material is considered an insulator.

Intrinsic semiconductors show a high state of purity and are perfectly crystallized.
Silicon is one of the most employed semiconductors for electronic device fabrication.
In the crystalline state, silicon atoms occupy tetrahedral positions in a face centred cubic
lattice similar to that of diamond, sharing their four valence electrons with the four nearest
neighbours in covalent bonds (Figure 3.4(a)). At a temperature of 0 K, all the bonds are
occupied by electrons and the valence band is completely full, whereas the conduction
band is completely empty. As the temperature increases over 0 K (Figure 3.4(b)), some of
the electrons can gain enough energy from the vibrations of the atoms (thermal energy) to
break a bond and become free; in this process a hole is also created. The energy needed
for the electron to make the transition between the valence band to the conduction band is
at least equal to the gap energy. It is important to remark that in intrinsic semiconductors
every time an electron is transferred from the valence band to the conduction band, a hole

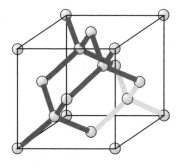

(a)

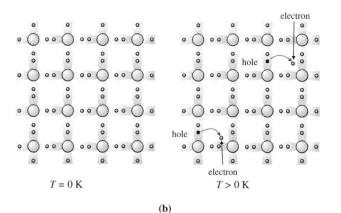

(b)

Figure 3.4. (a) Silicon lattice; (b) if $T > 0$ some of the bonds are broken liberating an electron and a hole. (The created electrons and holes can move freely through the crystal.)

is created in the valence band. Therefore, in an intrinsic semiconductor, the concentration of electrons should be equal to the concentration of holes. Evidently, photons with energies larger than the gap can also create electron–hole pairs, a phenomenon which is exploited in many optoelectronic devices.

The semiconductors mostly used in the fabrication of devices such as diodes, transistors, solar cells, etc. are extrinsic semiconductors, which can be obtained from intrinsic semiconductors by adding dopants in a controlled fashion. The concentration of added dopants determines the electrical conductivity of the extrinsic semiconductors. Figure 3.5 shows a bidimensional representation of the silicon lattice, each atom bonded to the four nearest neighbours. Suppose that some pentavalent impurity atoms such as As, Sb, etc. are added in a small concentration; usual dopant concentrations range from one impurity

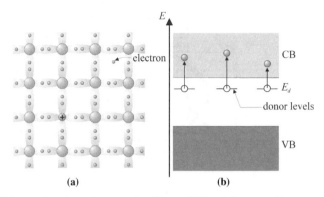

(a)   (b)

Figure 3.5. (a) Pentavalent impurity atom in silicon; (b) band diagram of a n-type semiconductor.
(The electrons can leave the impurity atoms if energy E_d is provided, and are free to
move through the crystal; the donor impurities become static positive ions.)

atom per 10^5 to 10^8 Si atoms. Figure 3.5(a) shows an impurity Sb atom of valence +5 in
a silicon lattice, where it substitutes a Si atom. Four of the valence electrons of Sb share
electrons with the four nearest neighbours. The fifth electron, however, does not participate in a covalent bond and remains very weakly bounded to the Sb atom. At ambient
temperature, most of these electrons get enough energy to leave the atom and become
free to move around the crystal. As a consequence, the impurity Sb atom becomes positively ionized. In an energy diagram, the above process can be represented by means
of the transfer of an electron from an energy level E_d below the conduction band to the
conduction band edge (Figure 3.5(b)). This type of impurities which yield electrons to the
conduction band are called *donors* and the corresponding semiconductor is said to be of
n-type, since the concentration of electrons significantly exceeds that of holes, i.e. $n \gg p$.
The ionization energy E_d can be estimated by comparing the above situation with that
of a hydrogenic atom, since the extra fifth electron is attracted by the Sb ion which is
positively charged. Therefore, Eq. (2.15) of Chapter 2 giving the energy of the hydrogen
atom can be used to estimate E_d after some corrections. First, since the Coulombic system
formed by the ion or positive charge and the electron is located within a material medium,
$(\varepsilon_r)^2$ should be included into the denominator where ε_r is the relative dielectric constant;
in addition, the electron mass in vacuum has to be substituted by the electron effective
mass in the semiconductor. This calculation yields values of E_d of about 0.05 eV, which
are of the order of the values experimentally determined.

   If trivalent impurities of valence +3 such as boron are added initially to intrinsic
silicon (Figure 3.6), the three valence electrons are covalently shared with three of
the four nearest neighbouring Si atoms; however, one of the otherwise covalent bond
has only one electron since there is a deficiency of one electron to complete the bond.

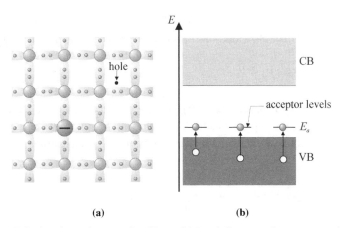

(a)   (b)

Figure 3.6. (a) Trivalent impurity atom in silicon; (b) band diagram of a p-type semiconductor. (If energy E_a is provided, the acceptor impurities can trap electrons from the valence band becoming static negative ions and simultaneously producing free holes.)

The state corresponding to the missing electron is therefore a hole (Section 2.6.3). If activated thermally, neighbouring electrons can move to occupy the empty state forming holes in the valence band. The activation energy for this process is usually very small and, therefore, at room temperature, most trivalent impurities have received the extra electron to complete the bond, becoming negative ions. In this process a number of holes similar to the number of impurities are created. Consequently, the addition of trivalent impurities, called acceptors, can create holes in the valence band. In an energy diagram the impurity levels can be represented at an acceptor ionization energy level, E_a, above the valence band edge, since electrons from the valence band are promoted in energy to the impurity acceptor level, thus creating a hole in the valence band.

Doping in III-V compounds can be discussed in similar terms to the case of silicon. If column VI impurities like Se or S occupy As (valence 5) sites in GaAs, they serve as donors. On the other hand, if column II impurities as for example Be or Cd occupy Ga (valence 3) sites, they act as acceptors. Column IV impurities (Si, Ge, etc.) added to III-V compounds, can serve either as donors or acceptors depending on whether they substitute atoms residing in column III or in column V sublattices, respectively. For instance, silicon in GaAs occupies Ga vacancies and therefore normally acts as a donor.

3.4. ELECTRON AND HOLE CONCENTRATIONS IN SEMICONDUCTORS

For the calculation of the concentration of carriers in a semiconductor, for instance electrons, it is necessary to previously know the density of electrons per energy interval in the conduction band, i.e. the density of states (DOS) function and the probability of each state being occupied. The distribution in energy of electrons, $n(E)$, is then given by

$$n(E) = \rho(E) f_{FD}(E) \tag{3.3}$$

where $\rho(E)$ is the DOS function and $f_{FD}(E)$ is given by the Fermi–Dirac distribution function. Substituting the expressions for $\rho(E)$ and $f_{FD}(E)$ given by Eqs (2.36) and (2.17) of Chapter 2, respectively, we get for the distribution of the electrons in energy in the conduction band:

$$n(E) = \frac{4\pi}{h^3} \left(2m_e^*\right)^{3/2} (E - E_c)^{1/2} \frac{1}{1 + e^{(E - E_F)/kT}} \tag{3.4}$$

since the lowest electron energy in the conduction band is E_c.

In order to obtain the concentration of electrons, n, in the conduction band, the distribution function $n(E)$ given by Eq. (3.4) should be integrated from the conduction band edge E_c to the highest level of the band $E_{c,max}$. However, since the Fermi level is located at a considerable energy below this level, the integral can be extended to ∞, making the calculations much easier. We have therefore to calculate the integral

$$n = \int_{E_c}^{\infty} n(E) \, dE \tag{3.5}$$

Substituting $n(E)$ by its expression of Eq. (3.4), we obtain:

$$n = N_c F_{1/2}(\alpha) \tag{3.6}$$

where the parameter α is given by

$$\alpha = \frac{E_F - E_c}{kT} \tag{3.7}$$

and N_c is the so-called *effective density of states* in the conduction band:

$$N_c = 2 \left(\frac{m_e^* kT}{2\pi \hbar^2} \right)^{3/2} \tag{3.8}$$

In Eq. (3.6), $F_{1/2}(\alpha)$ is the *Fermi integral* defined by

$$F_{1/2}(\alpha) = \frac{2}{(\pi)^{1/2}} \int_0^{\infty} \frac{y^2 dy}{1 + e^{(y - \alpha)}} \tag{3.9}$$

The Fermi integral does not have a closed analytical expression and therefore the electron concentration given by Eq. (3.6) has to be calculated by approximation methods. For $\alpha \leq -3$, the degree of approximation is excellent, and under this assumption

$$F_{1/2} \approx \exp\left(\frac{E_F - E_c}{kT}\right) \tag{3.10}$$

A semiconductor is called *non-degenerate* when the above approximation is fulfilled, i.e. if

$$E_c - E_F \geq 3kT \tag{3.11}$$

Under this condition, we can write with a high degree of approximation:

$$n = N_c e^{-(E_c - E_F)/(kT)} \tag{3.12}$$

for the electron concentration in the conduction band, where N_c is given by Eq. (3.8).

We can proceed in a similar way for the calculation of the hole concentration p in the valence band. The hole distribution function is given by

$$f_h(E) = 1 - f_{FD}(E) \tag{3.13}$$

since a hole represents an electron energy state which is vacant. Therefore, from Eq. (3.3)

$$p = \int_{-\infty}^{E_v} \rho_h(E)\, f_h(E)\, dE \tag{3.14}$$

where E_v is the valence band edge. Proceeding as we did for electrons, the hole concentration in a non-degenerate semiconductor is:

$$p = N_v\, e^{(E_F - E_v)/(kT)} \tag{3.15}$$

where

$$N_v \equiv 2\left(\frac{m_h^* kT}{2\pi\hbar^2}\right)^{3/2} \tag{3.16}$$

In this case N_v is called the effective density of states in the valence band. Figure 3.7 shows the DOS function (a), the Fermi–Dirac distribution function (b), and the distribution of carriers in energy (c), for intrinsic, n-type and p-type semiconductors, respectively.

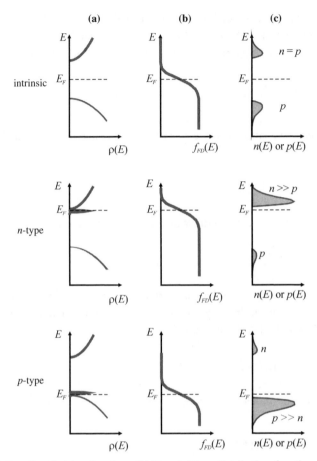

Figure 3.7. (a) Density of states function; (b) Fermi–Dirac distribution function; (c) distribution of carriers in energy. (The case of intrinsic n-type and p-type semiconductors are shown.)

The electron and hole concentrations given by Eqs (3.12) and (3.15) correspond to the shaded areas in the figure.

A very useful equation, called the *law of mass action* for charge carrier concentrations, can be derived from the above expressions for the electron and hole concentrations. In effect, from Eqs (3.12) and (3.15), we have:

$$np = N_c N_v \, e^{-(E_c - E_v)/(kT)} = N_c N_v e^{-E_g/kT} = \text{const } (T) \tag{3.17}$$

where E_g is the semiconductor gap. Since, in particular, the above equation has also to be verified for intrinsic semiconductors for which $n = p = n_i$, we can write Eq. (3.17) in terms of the carrier concentration n_i for the intrinsic material:

$$n_i = (np)^{1/2} = (N_c N_v)^{1/2} e^{-E_g/2kT} \tag{3.18}$$

From Eq. (3.18), we get the result that, for the case of intrinsic semiconductors, the carrier concentration and the electron–hole pair concentration decreases exponentially with the magnitude of the gap, and increases very strongly with temperature. As an example, for silicon at room temperature one obtains $n_i \approx 1.5 \times 10^{10} \, \text{cm}^{-3}$. This very low value of the carrier concentration (in a metal is of the order of $10^{22} \, \text{cm}^{-3}$) is the reason why most non-doped semiconductors at room temperature behave, from the point of view of electrical conduction, as if they were insulators.

In a given semiconductor, one usually knows the doping levels, i.e. the concentrations of donors N_D, and acceptors N_A, which are fixed during the fabrication of the crystals. The problem of finding the electron and hole concentrations is easy to solve if we are at temperatures high enough for all impurities to be ionized. This is indeed the case for the most semiconductors at room temperature. For instance, if $T > 100 \, \text{K}$ all the impurities in silicon are ionized, as a consequence of the low value of the impurity ionization energy (Section 3.3). In addition, since matter is in general neutral, the total positive charge, mobile and fixed, of the semiconductor should be equal to the total negative charge. Therefore:

$$p + N_D = n + N_A \tag{3.19}$$

which is the so-called *condition of charge neutrality*. From this equation and the law of mass action $np = n_i^2$, one immediately gets for the electron and hole concentrations, assuming that all impurities are ionized:

$$n = \frac{1}{2} \left[(N_D - N_A) + \sqrt{(N_D - N_A)^2 + 4n_i^2} \right] \tag{3.20}$$

$$p = \frac{1}{2} \left[(N_A - N_D) + \sqrt{(N_D - N_A)^2 + 4n_i^2} \right] \tag{3.21}$$

Observe that in the case in which $N_D = N_A$, the semiconductor behaves as intrinsic since the number of electrons and holes are equal, and therefore is called a *compensated semiconductor*.

For most extrinsic semiconductors at room temperature, one of the doping concentrations, N_D or N_A is much higher than n_i. In general, semiconductors are usually fabricated such that either $N_D \gg N_A$ or $N_A \gg N_D$. In the first case the semiconductor is n-type, since conduction is mainly due to electrons originated by the ionized donor impurities. For n-type semiconductors, we have from Eqs (3.20) and (3.21)

$$n \approx N_D, \quad p \approx \frac{n_i^2}{N_D} \tag{3.22}$$

Similarly, for p-type semiconductors:

$$p \approx N_A, \quad n \approx \frac{n_i^2}{N_A} \tag{3.23}$$

Evidently, the above approximations do not apply at high temperatures since n_i increases quasi-exponentially with T.

From Eqs (3.12) and (3.15), giving the carrier concentrations, it is also straightforward to calculate the Fermi level in some limiting cases. For *non-degenerate* n-type semiconductors, at temperatures high enough so that all impurities are ionized ($n \approx N_D$), but not so high that still $N_D \gg n_i$, the *Fermi level* from Eqs (3.12) and (3.22) is given by:

$$E_c - E_F = kT \ln \frac{N_c}{N_D} \tag{3.24}$$

and similarly for a p-type semiconductor:

$$E_F - E_v = kT \ln \frac{N_v}{N_A} \tag{3.25}$$

Observe from Eqs (3.24) and (3.25) that the higher the doping levels, the closer are the Fermi levels, located in the gap, to the respective bands edges.

If the semiconductor is intrinsic, $n = p$, and from Eqs (3.12) and (3.15), one gets for the Fermi level $(E_F)_i$:

$$(E_F)_i = \frac{E_c + E_v}{2} + kT \ln \frac{N_v}{N_c} = \frac{E_g}{2} + \frac{3}{4} kT \ln \frac{m_h^*}{m_e^*} \tag{3.26}$$

where E_g is the semiconductor gap. Observe that if T is relatively low, or if m_h^* and m_e^* have similar values, the Fermi level of intrinsic semiconductors is located close to the middle of the semiconductor gap.

3.5. ELEMENTARY TRANSPORT IN SEMICONDUCTORS

The mechanism of carrier transport in semiconductors, under the action of low electric fields, is quite similar to the case of metals. It is true that the electrical conductivity σ of semiconductors is much lower than for metals, but this can be mainly attributed to the much lower value of the carrier concentration, of the order of 10^{14} to 10^{17} cm^{-3}, in comparison to the case of metals (10^{22} cm^{-3}).

However, in semiconductors we have an additional type of conduction which does not have the equivalent in metals, and is due to the differences in carrier concentrations from region to region that often arise in semiconductors. One example can be a semiconductor fabricated with a non-uniform doping concentration. We can also change the carrier concentrations in a given region of the semiconductor by illumination with photons of enough energy to create electron–hole pairs. In such situations, electrons and holes move independently by diffusion due to their respective concentration gradients.

3.5.1. *Electric field transport. Mobility*

In order to explain low field electrical conduction in semiconductors, a semiclassical model can be considered. According to Eq. (2.57) of Section 2.6.1, if the force due to an electric field $\vec{F}$ is applied to an electron, its wave vector $\vec{k}$, and consequently its crystal momentum, would increase indefinitely. As we can imagine, this is not the case, since carriers, as they move by the action of the field, experience scattering events or collisions due to the existence of phonons, doping impurities, etc. In a semiclassical formulation, the scattering events are equivalent to a kind of frictional force acting in a direction opposite to the motion. Therefore, we should have according to Newton's second law in one dimension:

$$\left(qF - \frac{m_e^*}{\tau_e} \right) v_e = m_e^* \frac{dv_e}{dt} \tag{3.27}$$

for an electron of effective mass m_e^*. We have assumed in Eq. (3.27) that the frictional force is proportional to the electron drift velocity v_e. The parameter τ_e is known as the *relaxation time*, since when we disconnect the field:

$$v_e = [v_e]_{t=0}\, e^{-t/\tau_e} \tag{3.28}$$

i.e. the electron velocity acquired by the field goes exponentially to zero with a characteristic time equal to τ_e. From Eq. (3.27) we can appreciate that in the steady state, i.e. when the total external force acting on the electron is zero, then the drift velocity acquires a constant value given by:

$$v_e = -\frac{q\tau_e}{m_e^*} F \qquad (3.29)$$

The constant of proportionality between the electrical field and the drift velocity is called the electron *mobility* μ_e, i.e.:

$$\mu_e = \left| \frac{F}{v_e} \right| \qquad (3.30)$$

From the above definition, we can consider the mobility as being equal to the velocity acquired by the electron per unit electric field. According to Eqs (3.29) and (3.30), the electron mobility can also be written as:

$$\mu_e = \frac{q\tau_e}{m_e^*} \qquad (3.31)$$

Similarly, for the hole mobility, we have:

$$\mu_h = \frac{q\tau_h}{m_h^*} \qquad (3.32)$$

The electron current density can be expressed as:

$$J_e = -qnv_e \qquad (3.33)$$

where q is the electronic charge and n the carrier concentration, or by using Eq. (3.30) for the mobility:

$$J_e = qn\mu_e F \qquad (3.34)$$

Let us assume Ohm's law for the relation between J_e and E, i.e.

$$J_e = \sigma F \qquad (3.35)$$

where σ is the electrical conductivity. Since the current, and therefore the conductivity, can be due to both electrons and holes, we have from Eqs (3.33) and (3.35) and similar equations for holes, the following expression for the *electrical conductivity* in a semiconductor:

$$\sigma = q(n\mu_e + p\mu_h) \tag{3.36}$$

where n and p are the electron and hole concentrations, respectively.

It is interesting to observe that the simple model that we have introduced allows us to relate macroscopic parameters, such as electrical conductivity or mobility, with intrinsic properties of the carriers, like the effective mass. For instance, in the case of GaAs, m_e^* has a very low value ($m_e^* = 0.066m_0$) and therefore the electron mobility given by Eq. (3.31) is very large.

3.5.2. Conduction by diffusion

As we have previously mentioned, diffusion conduction in semiconductors is produced by gradients in carrier concentration. In reality, diffusion is originated by the random motion of the carriers and does not have anything to do with their charge. Carriers diffuse as a consequence of several factors: concentration gradients, the random thermal motion, and the scattering events produced by the lattice imperfections. Therefore, the diffusion current in a semiconductor obeys the general equation of diffusion. In one dimension, the diffusion currents for electrons and holes are given, respectively, by

$$J_e = q D_e \frac{dn}{dx} \tag{3.37}$$

and

$$J_h = -q D_h \frac{dp}{dx} \tag{3.38}$$

where D_e and D_h are known as the *diffusion coefficients* for electrons and holes, and q is the electronic charge. The negative sign of Eq. (3.38) is motivated by the negative sign of the derivative (areas of high carrier concentration to the left of those with low carrier concentration) and the fact that positive charges (holes) move in the positive x-direction under the action of the concentration gradient, which makes the current positive.

In a given semiconductor, the carriers are scattered as they move, either by the action of the electric field or the concentration gradient, by the same dispersion mechanisms. Therefore, the mobility and the diffusion coefficient cannot be independent. In fact, they are related by the *Einstein relations* for the electrons

$$\frac{D_e}{\mu_e} = \frac{kT}{q} \tag{3.39}$$

and for holes

$$\frac{D_h}{\mu_h} = \frac{kT}{q} \tag{3.40}$$

At room temperature D/μ equals 0.026 V.

3.5.3.   Continuity equations. Carrier lifetime and diffusion length

We have seen in previous sections how carriers can be generated in semiconductors. As an example we considered the generation of electron–hole pairs, either thermally or by incident photons. In addition, we have studied in Sections 3.5.1 and 3.5.2 the influence of an electric field and a gradient concentration on the carrier transport. Once the carriers are generated, they can recombine; for instance, an electron in the conduction band might fall to an empty energy state (hole) in the valence band. If we call g and r the *generation and recombination* rates, respectively, we can write a *continuity equation*, as it is done in electricity, for the rate of change of the electron density n. Let us assume for simplicity a one-dimensional model and consider that the concentration gradient is a function of time and position. The rate of change of the electron density in a small region between x and $x + dx$ is then given by

$$\frac{\partial n}{\partial t} = g - r + \mu_e \frac{\partial (nF)}{\partial x} + D_e \frac{\partial^2 n}{\partial x^2} \tag{3.41}$$

where the last two terms are the divergence ones obtained from Eqs (3.34) and (3.37). Evidently, a similar equation can be obtained for the rate of change of holes. Since both equations are coupled, the solution to them is quite complicated, unless we make some simplifying assumptions. For this, let us first define the *excess carrier concentrations*. If the electron concentration in equilibrium is n_0, the excess carrier concentration Δn is defined as:

$$\Delta n = n(x,t) - n_0 \tag{3.42}$$

where $n(x, t)$ is the actual electron concentration. Similarly, for holes we define the excess carrier concentration:

$$\Delta p = p(x,t) - p_0 \tag{3.43}$$

The first simplifying assumption is the charge balance or *neutrality condition* which assumes that the excess electron concentration is balanced by the one corresponding to holes, i.e.

$$\Delta n = n - n_0 = p - p_0 = \Delta p \qquad (3.44)$$

It is also assumed that either excess carrier concentration is much smaller than the larger of the two equilibrium concentrations, n_0 and p_0. Under these assumptions, it can be proved, in the case of strong character extrinsic semiconductors, that it is only necessary to solve the continuity equations for the minority carriers [1]. Therefore in a p-type semiconductor, we have for the excess electron concentration Δn (minority carriers):

$$\frac{\partial(\Delta n)}{\partial t} = g - r + \mu_e F \frac{\partial(\Delta n)}{\partial x} + D_e \frac{\partial^2(\Delta n)}{\partial x^2} \qquad (3.45)$$

Evidently, in the case of n-type semiconductors, the continuity equation states that:

$$\frac{\partial(\Delta p)}{\partial t} = g - r + \mu_h F \frac{\partial(\Delta p)}{\partial x} + D_h \frac{\partial^2(\Delta p)}{\partial x^2} \qquad (3.46)$$

where Δp is the excess hole concentration. Notice that in Eqs (3.45) and (3.46), F is the applied electrical field which we have assumed constant.

In semiconductor electronic devices such as p–n junctions, carriers of one sign, for instance electrons from an n-type semiconductor, cross the interface and enter a p-region, becoming minority carriers. In fact, the characteristics of electronic devices depend markedly on the behaviour of the injected minority carriers, which can be described by several parameters as their lifetime, diffusion length, etc. Let us suppose that an n-type semiconductor is illuminated by photons which create electron–hole pairs with steady excess carrier concentrations as defined in the previous section, such that $\Delta n = \Delta p \ll n_0$, $\Delta n = \Delta p \gg p_0$, where n_0 and p_0 are the equilibrium carrier concentrations in the n-type material. If at instant $t = 0$, we stop illuminating the material, the minority hole excess carrier concentration Δp decreases with time by recombination of the holes proportionally to its instant value of the concentration, i.e.:

$$\left[\frac{\partial \Delta p}{\partial t}\right]_{recomb} = -\frac{\Delta p}{\tau_h} \qquad (3.47)$$

where $1/\tau_h$ is the constant of proportionality. If we call $(\Delta p)_0$ the hole excess concentration at instant $t = 0$, we get after integrating:

$$\Delta p(t) = (\Delta p)_0 \, e^{-t/\tau_h} \qquad (3.48)$$

Evidently, τ_h can be interpreted as the *minority carrier lifetime* of holes in the n-type semiconductor. In the same way, a minority lifetime for electrons can be defined. Minority carrier lifetimes in doped semiconductors are of the order of 10^{-7}s, but they can be increased by the addition of impurities or traps to the semiconductor.

Suppose, as in Figure 3.8(a), a different hypothetical experiment in which carriers of one type, for instance electrons, are injected steadily into a p-type semiconductor at its surface ($x = 0$) and we want to find how the carrier concentration diminishes with distance x inside the p-type semiconductor, as a consequence of recombination with holes. For this, let us apply the continuity equation under steady state conditions and zero applied electric field. Under these conditions, Eq. (3.45) reduces to:

$$-\frac{\Delta n(x)}{\tau_e} + D_e\frac{\partial^2(\Delta n(x))}{\partial x^2} = 0 \qquad (3.49)$$

where we have made use of the definition, just introduced, of minority carrier lifetime. If we call N_0 the steady rate of injected electrons at $x = 0$, we get after integration:

$$\Delta n(x) = N_0\, e^{-x/L_e} \qquad (3.50)$$

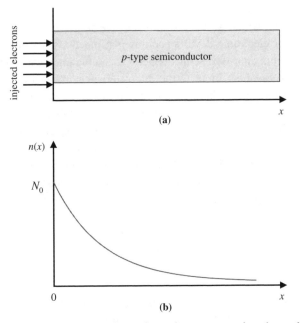

Figure 3.8. (a) Injection of electrons at the surface of a p-type semiconductor located at $x = 0$; (b) dependence of concentration of injected electrons as a function of the distance to the surface.

where

$$L_e = (D_e \tau_e)^{1/2} \tag{3.51}$$

Figure 3.8(b) represents how the concentration of injected electrons diminishes with distance inside the p-type semiconductor. The parameter L_e is known as the minority *diffusion length* for electrons and represents the distance for which the injected electrons diminish their concentration to $1/e$ of its value at $x = 0$.

3.6. DEGENERATE SEMICONDUCTORS

In Section 3.4 we gave the name non-degenerate semiconductors to those for which the Fermi level E_F is located in the gap at an energy of about $3kT$ or more away from the band edges. Since for these semiconductors, classical statistics could be applied, we derived simple expressions, Eqs (3.12) and (3.15), for the concentration of electrons and holes, respectively. Under these premises we also derived Eqs (3.24), (3.25), and (3.26) which give the location of the Fermi level for n-type, p-type, and intrinsic semiconductors, respectively.

As the dopant concentration is increased, the Fermi level approaches the band edges and when n or p exceeds N_c or N_v, given by Eqs (3.8) and (3.16), respectively, the Fermi level enters the conduction band in the case of n-type semiconductors or the valence band if the semiconductor is p-type. These heavily doped semiconductors are called *degenerate semiconductors* and the dopant concentration is usually in the range of 10^{19}–10^{20} cm^{-3}.

In the case of degenerate semiconductors, the wave functions of electrons in the neighbourhood of impurity atoms overlap and, as it happens in the case of electrons in crystals, the discrete impurity levels form narrow *impurity bands* as shown in Figure 3.9(a). The impurity bands corresponding to the original donor and acceptor levels overlap with the conduction and valence bands, respectively, becoming part of them. These states which are added to the conduction or valence bands are called *bandtail states*. Evidently, as a consequence of bandtailing, the phenomenon of *bandgap narrowing* is produced. Bandgap narrowing has important consequences in the operation of laser diodes (Section 10.3) and in the absorption spectrum of heavily doped semiconductors.

Figure 3.9(b) shows the energy diagram of a degenerate n-type semiconductor. As we know, the energy states below E_{Fn} are mostly filled. Therefore most of the electrons have energies in the narrow range between E_c and E_{Fn}. The band diagram is somewhat similar to metals and the Fermi level coincides with the highest energy of the electrons in the band. However, if one dopes very heavily, to about 10^{20} cm^{-3}, a carrier saturation effect appears as a consequence of interaction between dopants. For this very high concentration regime, Eqs (3.22), (3.23), and the law of mass action expressed in Eq. (3.17) do not apply.

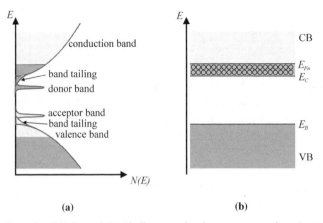

Figure 3.9. (a) Impurity bands and bandtail states in degenerate semiconductors; (b) energy diagram of an n-type degenerate semiconductor.

3.7. OPTICAL PROPERTIES OF SEMICONDUCTORS

3.7.1. *Optical processes in semiconductors*

When light incides on a semiconductor, we can observe a series of optical phenomena like absorption, transmission, and reflection. All these phenomena induce a series of electronic processes in the semiconductor, which can be studied by recording their respective optical spectra. The absorption spectrum of a typical semiconductor shows several significant features. The dominant absorption process occurs when the energy of the incident photons is equal or larger than the semiconductor gap and therefore electronic transitions from occupied valence band states to empty conduction band states become dominant. These transitions can be either direct or indirect and the absorption coefficient is calculated by means of time-dependent perturbation theory.

At the low energy side of the fundamental edge, exciton absorption can be observed as a series of sharp peaks. An exciton consists of a bound electron–hole pair in which the electron and hole are attracted by the Coulomb interaction and their absorption spectrum is studied in Section 3.7.3. Other absorption processes in semiconductors correspond to electronic transitions between donor levels and the conduction band states, and from the valence band to acceptor levels. The corresponding absorption peaks are located in the infrared ranges, as a consequence of the low values of the donor and acceptor ionization energies (Section 3.3). In heavily doped semiconductors, optical absorption by free carriers can also become significant, since the absorption coefficient is proportional to the carrier concentration. Finally, in ionic crystals, optical phonons can be directly excited by electromagnetic waves, due to the strong electric dipole coupling between photons and transverse optical phonons. These absorption peaks, arising from the lattice vibrations, appear in the infrared energy range.

The absorption of light by a semiconductor can be described macroscopically in terms of the absorption coefficient. If light of intensity I_0 penetrates the surface of a solid, then the intensity $I(z)$ at a distance z from the surface varies as

$$I(z) = I_0 e^{-\alpha z} \tag{3.52}$$

where α is a material property called the *absorption coefficient*, which depends on the light wavelength and is given in units of cm^{-1}. The parameter $1/\alpha$ is called the *penetration depth*. Evidently, the higher the absorption coefficient, the smaller the depth at which light can penetrate inside the solid. For semiconductors like GaAs, α increases very sharply when the photon energy surpasses E_g, since the optical transitions are direct (Section 3.7.2). On the contrary, for indirect semiconductors like Si or Ge, the increase of α is slower since the optical transitions require the participation of phonons. Therefore, the increase of α is not as sharp as in the case of direct semiconductors. In addition, the onset of absorption does not occur exactly when $h\nu = E_g$ as for direct transitions, but in an interval of the order of the energy of the phonons around E_g.

3.7.2. Interband absorption

Interband absorption across the semiconductor gap is strongly dependent on the band structure of the solid, especially on the direct or indirect character of the gap. In this section we are mainly going to review the case of interband optical transitions in direct gap semiconductors, since most of the optoelectronic devices of interest in light emission (Chapter 10) are based on this type of materials.

For the calculation of the optical absorption coefficient we have to make use of the quantum mechanical transition rate, W_{if}, between electrons in an initial state ψ_i which are excited to a final state ψ_f. This rate is given by the Fermi Golden rule of Eq. (2.26) of Chapter 2:

$$W_{if} = \frac{2\pi}{\hbar} \rho(E) \left| H'_{if} \right|^2 \tag{3.53}$$

In this expression, the matrix element H'_{if} corresponds to the optical external perturbation on the electrons and $\rho(E)$ is the density of states function for differences in energy E between final and initial states equal to the excitation photon energy $\hbar\omega$.

The perturbation Hamiltonian H' (Section 2.2.4) associated to electromagnetic waves acting on electrons of position vector $\vec{r}$ is given by

$$H' = -\vec{p} \cdot \vec{F} = e\vec{r} \cdot \vec{F} \tag{3.54}$$

and corresponds to the energy of the electric dipole of the electron $-e\vec{r}$ under the action of the wave electric field $\vec{F}$. Since the electronic states are described by Bloch functions (Section 2.4), we have to calculate matrix elements of the form

$$H'_{if} \propto \int_{u.c.} u_i^*(\vec{r}) x u_f(\vec{r}) d\vec{r} \qquad (3.55)$$

The integral in Eq. (3.55) is extended to the volume of the unit cell, since the integral over the whole crystal can be decomposed as a sum over the unit cells. The functions $u_i(\vec{r})$ and $u_f(\vec{r})$ in Eq. (3.55) have the periodicity of the lattice, according to the Bloch theorem.

The density of states function $\rho(E)$ that appears in Eq. (3.53) has to be calculated at the energy $h\nu$ of the incident photons, since the final and initial states, which lie in different energy bands, should be separated in energy by $h\nu$. For this reason the function $\rho(E)$ is usually called the *optical joint density of states function*, and for its calculation one has to know the structure of the bands. Another condition which has to be fulfilled is that, for direct gap semiconductors, the electron wave vector of the final state should be the same as the one of the initial state, i.e.

$$\vec{k}_f = \vec{k}_i \qquad (3.56)$$

since the momentum associated to the photon can be considered negligible.

Optical transitions around $\vec{k} = 0$ in direct III-V semiconductors like GaAs involve the valence band of p-like atomic orbitals and the conduction band originated from s-like orbitals. It is also known from the electric dipole selection rules that transitions from p-states to s-states are allowed, and therefore a strong optical absorption should be expected.

Let us now calculate the optical DOS. We can observe that the conduction band as well as the three valence bands (heavy hole, light hole, and split-off band) are all parabolic close to the Γ point (Figure 3.2(a)). The direct gap E_g equals the energy difference between the minimum of the conduction band and the maxima of the heavy and light hole bands, which are degenerated at Γ. For these transitions, conservation of energy requires

$$h\nu = E_g + \frac{\hbar^2 k^2}{2m_e^*} + \frac{\hbar^2 k^2}{2m_h^*} = E_g + \frac{\hbar^2 k^2}{2\mu} \qquad (3.57)$$

where μ is the reduced mass of the electron–hole system, and we have taken into account only one of the two degenerated hole bands for simplicity; the split-off hole band would show the absorption transition at higher photon energies. Considering the expression found in Section 2.3, Eq. (2.36), for the DOS function, we should have the absorption coefficient α for $h\nu \geq E_g$:

$$\alpha(h\nu) = \frac{1}{2\pi^2} \left(\frac{2\mu}{\hbar^2}\right)^{3/2} \left(h\nu - E_g\right)^{1/2} \tag{3.58}$$

Taking into account the expression for the rate of optical transitions W_{if} given by Eq. (3.53) and that the optical absorption coefficient α is proportional to W_{if}, then, for $h\nu \geq E_g$, we find that α should be also proportional to the square root of the photon energy minus the bandgap. Therefore, a plot of α^2 as a function of $h\nu$ should yield a straight line which intercepts the horizontal axis ($\alpha = 0$) at a value of the energy equal to the semiconductor gap.

As an example, we show in Figure 3.10, α^2 as a function of the photon energy for PbS which shows the linear behaviour just discussed [2]. Some other direct gap semiconductors do not verify this relation so exactly, since often not all the assumptions that we have considered in the derivation of Eq. (3.58) are fulfilled. For instance, in the case of GaAs at low temperatures, the region around the onset of this absorption can be overshadowed by the exciton absorption that will be studied in the next section. Also, Eq. (3.58) was only strictly valid for values of k close to $\vec{k} = 0$, but, as the photon energy increases, this condition does not hold anymore.

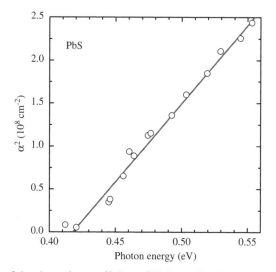

Figure 3.10. Square of the absorption coefficient of PbS as a function of photon energy. After [2].

3.7.3. *Excitonic effects*

As we have seen, photons of energy larger than the semiconductor gap can create electron–hole pairs. Usually the created electron and hole move independently of each other, but in some cases, due to the Coulomb interaction between them, the electron and hole can remain together forming a new neutral particle called *exciton*. Since excitons have no charge, they cannot contribute to electrical conduction. Exciton formation is very much facilitated in quantum well structures (Sections 1.5 and 4.10), because of the confinement effects which enlarge the overlapping of the electron and hole wave functions.

The simplest picture of an exciton consists of an electron and a hole orbiting inside the lattice around their centre of mass, as a consequence of the Coulombic attraction between them (Figure 3.11). There are two basic types of excitons: (a) Excitons for which the wave function of the electron and hole have only a slight overlap, i.e. the exciton radius encompasses many crystal atoms. These excitons are called *Wannier–Mott excitons* and are usually detected in semiconductors. (b) Other excitons, mainly observed in insulators, have a small radius of the order of the lattice constant and are called *Frenkel excitons*.

Wannier–Mott excitons can be described according to a model similar to the hydrogen atom. Considering the exciton as a hydrogenic system, the energies of the bound states should be given by an expression similar to Eq. (2.15) of Chapter 2, with the proper corrections, in a similar manner as we did with ionization donor and acceptor impurity levels (Section 3.3) in extrinsic semiconductors. Evidently, the mass in this expression should be now the reduced mass μ of the system formed by the electron and hole effective

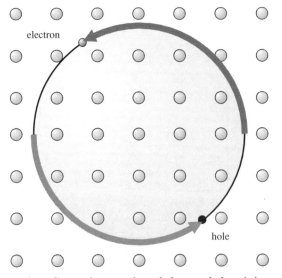

Figure 3.11. Representation of an exciton as a bound electron–hole pair in semiconductors.

masses, i.e.

$$\frac{1}{\mu} = \frac{1}{m_e^*} + \frac{1}{m_h^*} \tag{3.59}$$

In addition, we have to consider that the electron and the hole are immersed within a medium of dielectric constant $\varepsilon_r \varepsilon_0$, where ε_r is the high frequency relative dielectric constant of the medium. The bound states of the excitons are, therefore, given by:

$$E_n = -\frac{\mu R_H}{m_0 \varepsilon_r^2} \frac{1}{n^2} = -\frac{\mu}{m_0 \varepsilon_r^2} \frac{13.6\,\text{eV}}{n^2} = -\frac{R_{ex}}{n^2}, \quad n = 1, 2, 3, \dots \tag{3.60}$$

where R_H is the Rydberg constant for the hydrogen atom and R_{ex} is called the exciton Rydberg constant.

Figure 3.12 shows the excitonic bound states given by Eq. (3.60) and the exciton ionization energy E_I. The energy needed for a photon to create an exciton is smaller than the one needed to create an independent electron–hole pair, since we can think of this process as creating first the exciton and, later, separating the electron from the hole by providing an amount of energy equal to the exciton binding energy. Therefore, as shown in Figure 3.12, the exciton bound states, given by Eq. (3.60), should be located within the gap just below the edge of the conduction band.

Figure 3.13 shows the absorption spectrum of GaAs for photon energies close to the gap [3]. It is seen that the first three peaks predicted by Eq. (3.60), with $R_{ex} = 4.2\,\text{meV}$, are well resolved. This is because the spectrum was taken at very low temperatures, the spectrometer had a high resolution, and the sample was ultrapure. The advantage of

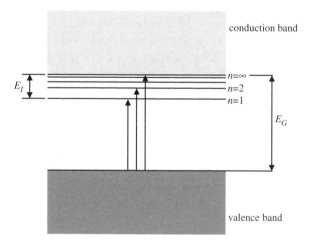

Figure 3.12. Excitonic states located in the gap, close to the conduction band edge.

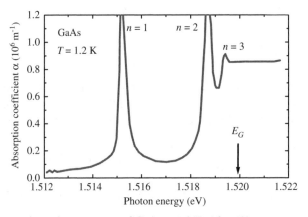

Figure 3.13. Exciton absorption spectrum of GaAs at 1.2 K. After [3].

working at low temperatures is double. On the one hand, excitons are less likely to be destroyed by phonons and, on the other, the thermal broadening of the absorption lines is reduced. Excitons are much better observed in intrinsic semiconductors than in doped ones, where the free charge carriers partially screen the Coulombic interaction between the electron and the hole.

A closer look to the band structure of semiconductors allows predicting which regions in k-space are favourable for the formation of excitons. Since the exciton is composed of a bound electron–hole pair, the velocity vectors of both particles should be the same, and therefore, according to Eq. (2.52) of Chapter 2, their respective conduction and valence bands should be parallel. This is evidently the case in the vicinity of the point $\vec{k} = 0$ in GaAs, i.e. around the spectral region corresponding to the direct gap.

If the intensity of the light that creates excitons is high enough, their density increases so much that they start to interact among themselves and with the free carriers. In this high density regime, *biexcitons* consisting of two excitons can be created. Biexcitons have been detected in bulk semiconductors as well as in quantum wells and dots. Biexcitons consist of two electrons and two holes and, similarly to the way followed to study excitons in terms of hydrogenic atoms, they can be compared to hydrogen molecules. In addition to biexcitons, *trions* consisting of an exciton plus either a hole or an electron have also been experimentally detected in several nanostructures, among them, III-V quantum wells and superlattices.

3.7.4. Emission spectrum

In Section 3.7.2, we have considered transitions of electrons from the valence band to the conduction band in semiconductors caused by photon absorption. In the inverse process, light can be emitted when an excited electron drops to a state in a lower energy band. If in this process light is emitted, we have *photoluminescence* due to a *radiative transition*. The emitted photons have in general a different frequency than the previously absorbed, and the emission spectrum is usually much narrower than the absorption spectrum. In effect, suppose, as in Figure 3.14, that a photon of energy $h\nu > E_g$ is absorbed and, as a consequence, an electron–hole pair is produced. In this process, the electron and/or the hole can get an energy higher than the one corresponding to thermal equilibrium. Subsequently, the electrons (the same applies to holes) lose the extra kinetic energy very rapidly by the emission of phonons (mainly optical phonons) occupying states close to the bottom of the band. The time taken in this step is as short as 10^{-13}s due to the strong electron–phonon coupling. On the other hand, the lifetime of the radiative process by which electrons drop to the valence band, by the emission of photons of $h\nu \sim E_g$, is several orders of magnitude longer (of the order of nanoseconds). Therefore, the emission spectrum should range between E_g and E_g plus an energy of the order of kT, since the electrons have enough time to get thermalized at the bottom of the conduction band. At present, these phenomena can be studied very nicely by means of very fast time-resolved photoluminescence spectroscopy, using ultra short laser pulses. Figure 3.15 shows the spectra of bulk GaAs at 77 K, after having been excited by 14 fs laser pulses [4]. The curves are shown for three different carrier concentrations of increasing values from top to bottom. The fourth curve in the figure represents the autocorrelation (AC) of the laser pulse.

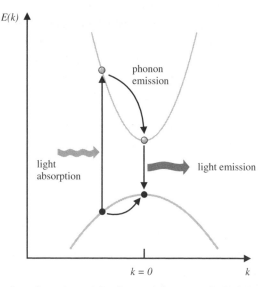

Figure 3.14. De-excitation of an electron by first emitting an optical phonon and subsequently a photon with energy approximately equal to the gap energy.

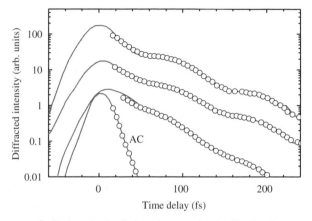

Figure 3.15. Spectra of GaAs obtained by time-resolved photoluminescence spectroscopy. After [4].

For many semiconductors, recombination of electrons and holes is mainly non-radiative, that is, instead of emitting photons, the recombination process occurs via *recombination centres* with energy levels located within the gap. In this case, the energy lost by the electrons is transferred as heat to the lattice. We have already seen in Section 3.3 that impurity states in semiconductors are located within the gap, close to the band edges, but other defects, such as vacancies or metal impurities, can have their levels much deeper within the gap. Only for quantum wells of very high quality, the number of emitted photons divided by the number of excited electron–hole pairs reaches values in the range 0.1–1. Even in direct gap bulk semiconductors, the values of the *luminescence yield* are very low, between 10^{-3} and 10^{-1}. Impurities and in general any kind of defects can act as recombination centres by first capturing an electron or a hole and subsequently the oppositely charged carrier. The defects which make possible electron–hole non-radiative recombination are colloquially called *traps* and in them the recombination centre re-emits the first captured carrier before capturing the second carrier. The recombination centres are called either fast or slow depending on the time that the first carrier remains at the centre before the second carrier is captured.

3.7.5. *Stimulated emission*

Suppose a simple electron system (Figure 3.16) of just two energy levels E_1 and E_2 ($E_2 > E_1$). Electrons in the ground state E_1 can jump to the excited state E_2 if they absorb photons of energy $E_2 - E_1$. On the contrary, photons of energy $E_2 - E_1$ are emitted when the electron drops from E_2 to E_1. In general, the emission of light by a transition from the excited state E_2 to the ground state E_1, is proportional to the population of

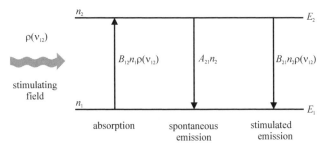

Figure 3.16. (a) Absorption; (b) spontaneous emission; (c) stimulated emission processes under steady state conditions. (The absorption should be equal to the sum of spontaneous and stimulated emission processes.)

electrons n_2 at the level E_2. This is called *spontaneous emission* and the proportionality coefficient is called the Einstein A_{21} coefficient, while the corresponding one for the *absorption* process is called the Einstein B_{12} coefficient. As Einstein observed, electrons can also drop from E_2 to E_1 if they are stimulated by photons of energy $h\nu = E_2 - E_1$. This process is therefore called *stimulated emission* and is governed by the Einstein B_{21} coefficient. Since stimulated emission is proportional to the density $\rho(\nu)$ of photons, in order to have a high rate of stimulated emission, in comparison to the spontaneous one, the radiation energy density should be very high. Evidently the above three Einstein coefficients are related to each other, since in the steady state the rate of upward and downward transitions shown in Figure 3.16 should be equal.

One very interesting aspect of stimulated emission is that the emitted photons are in phase with the stimulating ones. Precisely, the operation of lasers is based on the process of stimulated emission. Semiconductor lasers, which will be studied in Chapter 10, produce monochromatic and coherent light. The rate of stimulated emission should be proportional to $n_2\rho(\nu)$ and therefore, in order to dominate over absorption (proportional to n_1) we should have $n_2 > n_1$. This condition is known as *population inversion*, since in thermal equilibrium, according to the Boltzmann distribution, we have $n_1 < n_2$. Observe also that since stimulated emission is proportional to the radiation energy density, lasers need to make use of *resonant cavities* in which the photon concentration is largely increased by multiple internal optical reflections.

Population inversion in semiconductor lasers is obtained by the injection of carriers (electrons and holes) across p^+–n^+ junctions of degenerate direct gap semiconductors, operated under forward bias. Figure 3.17(a) shows a non-biased p–n junction and Figure 3.17(b) shows the junction when it is polarized under a forward bias. In this situation, a region around the interface between the p^+ and n^+ materials, called the *active region*, is formed, in which the condition of population inversion is accomplished.

From Figure 3.18(a), it can be deduced the range of energies of incoming photons which are able to produce a rate of stimulated emission larger than the absorption rate, therefore resulting in an *optical gain*. Taking into account the density in energy of electrons through density of states functions for the conduction and valence bands, one can deduce qualitatively the dependence of optical gain as a function of the energy of the incident photons. Evidently, as indicated in Figure 3.18(b), the photons that induce stimulated emission should have energies larger than E_g and lower than $E_{Fn} - E_{Fp}$. At higher temperatures, the Fermi–Dirac distribution broadens around the Fermi levels and as a result there is a diminution in optical gain.

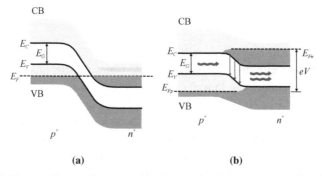

Figure 3.17. (a) Energy diagram for a p–n junction made of degenerate semiconductors with no bias; (b) idem, with a forward bias, high enough to produce population inversion in the active region.

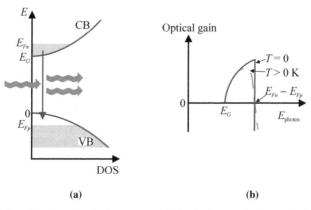

Figure 3.18. (a) Density of states of electrons and holes in the conduction and the valence bands, respectively; (b) optical gain as a function of photon energy.

REFERENCES

[1] McKelvey, J.P. (1966) *Solid-State and Semiconductor Physics* (Harper and Row, New York).

[2] Schoolar, R.B. & Dixon, J.R. (1965) *Phys. Rev.*, **137**, A667.

[3] Fehrenbach, G.W., Schäfer, W. & Ulbrich, R.G. (1985) *J. Luminescence*, **30**, 154.

[4] Banyai, L., Tran Thoai, D.B., Reitsamer, E., Haug, H., Steinbach, D., Wehner, M.U., Wegener, M., Marschner, T. & Stolz, W. (1995) *Phys. Rev. Lett.*, **75**, 2188.

FURTHER READING

Seeger, K. (1999) *Semiconductor Physics* (Springer, Berlin).

Singh, J. (2003) *Electronic and Optoelectronic Properties of Semiconductor Structures* (Cambridge University Press, Cambridge).

Yu, P. & Cardona, M. (1996) *Fundamentals of Semiconductors* (Springer, Berlin).

PROBLEMS

1. **Electron mean free path**. Find the electron mean path in GaAs, at room temperature and $T = 77$ K knowing that the respective mobilities are approximately 3×10^5 and 10^4 cm^2/Vs, respectively.

2. **Semiconductor doping**. In order to make a p–n diode, a sample of silicon of type n is doped with 5×10^{15} phosphorous atoms per cm^3. Part of the sample is additionally doped, type p, with 10^{17} boron atoms. (a) Calculate the position of the Fermi levels at $T = 300$ K, in both sides of the p–n junction. (b) What is the contact potential?

3. **Carrier concentrations in germanium**. Determine the free electron and hole concentration in a Ge sample at room temperature, given a donor concentration of 2.5×10^{14} cm^{-3} and acceptor concentration of 3.5×10^{14} cm^{-3}, assuming that all impurities are ionized. Determine its n or p character if the intrinsic carrier concentration of Ge at room temperature is $n_i = 2.5 \times 10^{13}$ cm^{-3}.

4. **Carrier concentrations in silicon**. A semiconducting silicon bar is doped with a concentration of 4×10^{14} cm^{-3} n-type impurities and 6×10^{14} cm^{-3} of p-type impurities. Assuming that the density of states is constant with increasing temperature and that electron mobility is twice that of holes, determine the carrier concentration and the conducting type at 300 and 600 K. The bandgap energy of silicon is $E_g = 1.1$ eV and n_i (300 K) $= 1.5 \times 10^{10}$ cm^{-3}.

5. **Diffusion currents in semiconductors**. The electron density in an n-type GaAs crystal varies following the relationship $n(x) = A \exp(-x/L)$, for $x > 0$, being

$A = 8 \times 10^{15}\,\text{cm}^{-3}$ and $L = 900\,\text{nm}$. Calculate the diffusion current density at $x = 0$ if the electron diffusion coefficient equals $190\,\text{cm}^2\text{s}^{-1}$.

6. **Diffusion length.** In a p-type GaAs sample electrons are injected from a contact. Considering the mobility of minority carriers to be $3700\,\text{cm}^2\text{V}^{-1}\text{s}^{-1}$, calculate the electron diffusion length at room temperature if the recombination time is $\tau_n = 0.6\,\text{ns}$.

7. **Carrier dynamics in semiconductors.** The electron energy close to the top of the valence band in a semiconductor can be described by the relationship $E(k) = -9 \times 10^{-36}k^2\,(J)$, where k is the wave vector. If an electron is removed from the state $\vec{k} = 2 \times 10^9 \vec{k}_u$, being $\vec{k}_u$ the unity vector in the x-direction, calculate for the resulting hole: (a) its effective mass, (b) its energy, (c) its momentum, and (d) its velocity. *Hint*: suppose that we are dealing with states close to the maximum of the valence band or to the minimum of the conduction band, thus being the parabolic dispersion relationship a good approximation.

8. **Energy bands in semiconductors.** The conduction band in a particular semiconductor can be described by the relationship $E_{cb}(k) = E_1 - E_2 \cos(ka)$ and the valence band by $E_{vb}(k) = E_3 - E_4 \sin^2(ka/2)$ where $E_3 < E_1 - E_2$ and $-\pi/a \leq k \leq +\pi/a$. Determine: (a) the bandgap, (b) the variation between the extrema $(E_{\max} - E_{\min})$ for the conduction and valence bands, (c) the effective mass for the electrons and holes in the bottom of the conduction band and the top of the valence band, respectively. Sketch the band structure.

9. **Excess carriers.** An n-type Ge bar is illuminated, causing the hole concentration to be multiplied by five. Determine the time necessary for the hole density to fall to $10^{11}\,\text{cm}^{-3}$ if $\tau_h = 2.5\,\text{ms}$. Assume the intrinsic carrier concentration to be $10^{13}\,\text{cm}^{-3}$ and the donor density $8 \times 10^{15}\,\text{cm}^{-3}$.

10. **Optical absorption in semiconductors.** A 700 nm thick silicon sample is illuminated with a 40 W monochromatic red light ($\lambda = 600\,\text{nm}$) source. Determine: (a) power absorbed by the semiconductor, (b) power dissipated as heat, and (c) number of photons emitted per second in recombination processes originated by the light source. The absorption coefficient at $\lambda = 600\,\text{nm}$ equals $\alpha = 7 \times 10^4\,\text{cm}^{-1}$ and the bandgap of Si is $1.12\,\text{eV}$.

11. **Excitons in GaAs.** (For the following problem take $\mu = 0.05m_0$ as the exciton reduced mass and $\varepsilon_r = 13$ as the relative dielectric constant.) (a) Calculate the Rydberg energy R_H. What is the largest binding energy? (b) Calculate the exciton radius a_{ex} in the ground level, following Bohr's theory. (c) Calculate the number of unit cells of GaAs ($a_0 = 0.56\,\text{nm}$) inside the exciton volume. (d) Supposing that the exciton is in its ground sate, up to what temperature is the exciton stable?

12. **Excitons in a magnetic field.** Suppose that a magnetic field B is applied to a gallium arsenide sample. Find the value of B for which the exciton cyclotron energy is equal to the exciton Rydberg energy of Eq. (3.60). Take $\varepsilon = 13$ for the value of the dielectric constant.

Chapter 4

The Physics of Low-Dimensional Semiconductors

Chapter 4

The Physics of Low-Dimensional Semiconductors

4.1. INTRODUCTION

During the last few decades, advances in solid state physics have been characterized by a gradual change in interest, from bulk crystals to solids which are very small in at least one of their three dimensions. Since they are easy to produce, initial focus mainly centred on very thin solid films, i.e. with a thickness comparable to the de Broglie wavelength λ_B of the electrons in the solid. From this research, a new class of effects arose, such as the quantum Hall effect (QHE), discovered by von Klitzing (1980) for which he received the Nobel Prize in 1985.

When one of the three spatial dimensions of a solid, usually a semiconductor material, is of a size comparable to λ_B, we say that we are dealing with a material or a structure of low dimensionality. In some situations, other characteristic lengths, different from λ_B, and revised in Section 1.3, are more convenient to use. For most important semiconductors, λ_B usually ranges between 10 and 100 nm and therefore we have to deal with solids of size in the nanometre range in order to observe quantum effects such as QHE, Coulomb blockade, quantized conductance, etc. (Chapters 6 and 7). We recall from Chapter 1 that low-dimensional materials are classified according to the number of spatial dimensions of nanometric size. If only one of the three dimensions is small enough, then the structure is called a *quantum well* (2D). In the case of *quantum wires* (1D), two of the dimensions are in the nanometric range, and finally in the case that the three dimensions are comparable to λ_B the structure is called a *quantum dot* (0D).

As seen in this chapter, a very thin semiconductor layer of nanometric size, e.g. GaAs, surrounded on each side by higher energy gap semiconductors such as AlGaAs, is frequently the main constituent of modern devices in optoelectronics. Another very interesting structure is simply formed by the junction, or more accurately heterojunction, of two semiconductors of different energy gaps (Section 1.7). In both cases, a potential well for electrons, similar to that in a MOS structure, is formed at the interface. If these wells have a width comparable to λ_B, then the electron energy levels are quantized in the well. These heterojunctions are, for example, the basis of the very fast MODFET transistors (Chapter 9).

4.2. BASIC PROPERTIES OF TWO-DIMENSIONAL SEMICONDUCTOR NANOSTRUCTURES

One of the most practical two-dimensional semiconductor structures consists of a sandwich of gallium arsenide (GaAs), with a thickness in the nanometre range, surrounded on each side by a semiconductor such as aluminium gallium arsenide ($Al_xGa_{1-x}As$) of higher bandgap. The bandgap of $Al_xGa_{1-x}As$ ($x \sim 0.3$) is close to 2.0 eV while that of GaAs is 1.4 eV. As a consequence, the potential energy profile has the shape of a square well (Figure 4.1(a)), with a barrier height of 0.4 eV for electrons and 0.2 eV for holes. In reality, the profile of the potential barrier is somewhat more complicated, since the potential varies with atomic distances, which also affects the wave functions. However, in most cases, it is a good approximation to consider the average over a few atomic distances (*envelope function approximation*). As seen in Figure 4.1(a), carrier motion for both electrons and holes is not allowed in the direction perpendicular to the well, usually taken as the z-direction because of the potential walls. However in the other two spatial directions (x, y), parallel to the semiconductor interfaces, the motion is not restricted, i.e. the electrons behave as free electrons.

The behaviour of electrons when their motion is restricted along one direction in the wells of infinite height corresponds to a well-known problem in quantum mechanics, the so-called particle in a box of infinite wells. In Section 4.3 we will address the problem for the case of barriers of finite height at the interfaces. It is well known from quantum mechanics that, in the case of infinite potentials barriers, the wave functions and energy levels of the bound electrons are given by

$$\psi_n(z) = \left(\frac{2}{a}\right)^{1/2} \sin\left(\frac{\pi n z}{a}\right) \tag{4.1}$$

$$E_n = \frac{\hbar^2 \pi^2}{2m_e^* a^2} n^2, \quad (n = 1, 2, \ldots) \tag{4.2}$$

where m_e^* is the effective mass of the electrons in the well material for the motion along the z-direction and a is the width of the well. From Eq. (4.2) we can derive several important consequences: (1) In general, quantum size effects will be more easily observable in quantum structures of very small size a, and for materials for which the electron effective mass is as small as possible. In this sense, GaAs nanostructures are very convenient since $m_e^* \sim 0.067m_0$, where m_0 is the free electron mass. This is equivalent to saying that in materials for which the electron mobility (Section 3.5.1) or the free electron path are large, quantum effects are easier to observe. (2) Quantum size effects, which require energy transitions of electrons between levels, are better observed at low temperatures, since the mean thermal energy of carriers is of the order of kT.

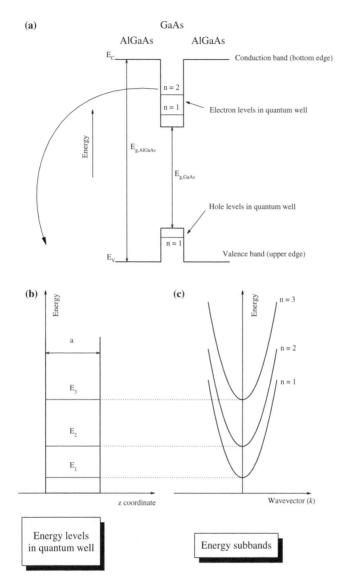

Figure 4.1. (a) AlGaAs–GaAs–AlGaAs square potential well; (b) energy levels; (c) energy subbands.

As it has been described, the motion of electrons in the quantum well is confined only in one direction, z, but in the (x, y) planes the electrons behave as in a three-dimensional solid. Therefore the electron wave function is separable as the product of ψ_x, ψ_y, and ψ_z, i.e.

$$\Psi = \psi_x \psi_y \psi_z \tag{4.3}$$

where, in our simple model, ψ_x and ψ_y satisfy the Schrödinger equation for a free electron, i.e. a travelling wave, while ψ_z is given by the Schrödinger equation with a square
well potential $V(z)$ and therefore can be expressed as in Eq. (4.1).

The expression for the total energy of electrons in the potential well, can then be
written as

$$E\left(k_x, k_y, n\right) = \frac{\hbar^2}{2m_e^*}\left(k_x^2 + k_y^2\right) + E_n = \frac{\hbar^2}{2m_e^*}\left(k_x^2 + k_y^2\right)$$

$$+ \frac{\hbar^2\pi^2}{2m_e^* a^2}n^2, \quad (n = 1, 2, \ldots) \tag{4.4}$$

where the quasi-continuous values of k_x, k_y are determined by the periodic boundary
conditions as in the case of a free electron in the bulk (Section 2.3).

Figure 4.1(b) schematically shows the discrete values of E_n corresponding to motion
in the z-direction while Figure 4.1(c) represents the E vs p dependence (remember that
$p = \hbar k$) for values of $\vec{p}$ in the plane (p_x, p_y). For each fixed value of E_n, the values of E
as a function of $\vec{p}$ form the so-called *energy subbands* represented in Figure 4.1(c). It is
interesting to observe that the lowest energy of electrons, E_1, is different from zero, in
contrast with classical mechanics. This result is expected from quantum mechanics since
a value of $E = 0$ would violate the uncertainty principle. In these systems the value of
$E = E_1$ is called zero-point energy.

From solid state physics we know that many physical properties, e.g. optical absorption,
transport of electronic current, etc. depend both on the energy spectrum and the density
of states (DOS) function, which gives the concentration of electrons at each value of the
energy. In a three-dimensional electronic system, it is already known (Section 2.3) that
the DOS depends as a parabolic function on the energy. However in the two-dimensional
case, this dependence is completely different. Proceeding as in the 3D case, it can be
appreciated that for the 2D case (Figure 4.2) the possible values of k_x, k_y are separated
by $2\pi/L$, where L is the dimension of the sample, which has been assumed to be square,
without loss of generality. The number of states in the k-space within a circular ring
limited by the circumferences of radii k and $k + dk$ is therefore:

$$n_{2D}(k)dk = \frac{2\pi k dk}{(2\pi/L)^2}$$

and the number of states in k-space per unit area is:

$$n_{2D}(k) = \frac{k}{2\pi} \tag{4.5}$$

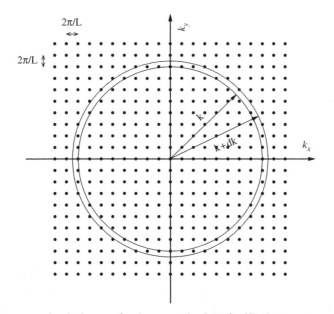

Figure 4.2. Representation in k-space for the states (k_x, k_y) of a 2D electron system.

If we wish to calculate the DOS in energy, we define $n_{2D}(E)$ such that $n_{2D}(E)\delta E$ is the number of states in the range δE. The densities of states in energy and wave vector are related by:

$$n_{2D}(E)\delta E = n_{2D}(k)\delta k \qquad (4.6)$$

where E and k are related by $E = \hbar^2 k^2 / 2m_e^*$. Differentiating this expression and taking into account Eq. (4.5), we have, after substitution in Eq. (4.6), and adding a factor 2 which accounts for the spin:

$$n_{2D}(E) = 2\frac{k}{2\pi}\frac{\delta k}{\delta E} = 2\frac{k}{2\pi}\frac{m_e^*}{\hbar^2 k} = \frac{m_e^*}{\pi\hbar^2} \qquad (4.7)$$

Note that in the 2D case, the DOS function is a constant, independent of energy. Let us show next that the DOS function, for the two-dimensional case, exhibits a staircase-shaped energy dependence (Figure 4.3) in which all the steps are of the same height, but located at energies E_n given by Eq. (4.2). In effect, from Figure 4.1(c) it can be appreciated that the interval of energy between 0 and E_1 is not allowed. For E such that $E_1 < E < E_2$ the electrons will be located in the subband corresponding to $n = 1$ and the value will be $m_e^*/\pi\hbar^2$. For the energy interval between E_2 and E_3, the electrons can be located

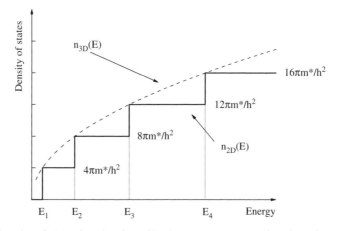

Figure 4.3. Density of states function for a 2D electron system, as a function of energy.

either in the $n = 1$ or in the $n = 2$ subbands, and consequently the DOS function would be twice the above value, i.e. $2m_e^*/\pi\hbar^2$, etc. The staircase shape of the $n_{2D}(E)$ function can be directly observed by optical absorption measurements, as we will see in Chapter 8. We have also represented the parabolic 3D case in Figure 4.3, from which it can be appreciated that the differences between the 2D and 3D cases are more discernable for low values of n.

According to the subband in which the electrons are located, their kinetic energy is differently partitioned as a consequence of the relationship expressed in Eq. (4.4). For instance, for a given energy in the interval E_2 to E_3, for the same value of energy, the electrons located in subband $n = 2$ have higher energy in the z-direction, E_2, than those in the $n = 1$ subband for which the corresponding value is E_1. Therefore the energy of motion corresponding to the plane (p_x, p_y) should be smaller for the electrons in subband $n = 2$. Evidently the separation of energy in different "components" (remember that the energy is a scalar function) is a direct consequence of the simple forms adopted for ψ and the energy in Eqs (4.3) and (4.4).

4.3. SQUARE QUANTUM WELL OF FINITE DEPTH

The quantum wells for electrons and holes in GaAs nanostructures surrounded by higher gap AlGaAs, studied in the previous section, are not of infinite height, as was assumed in order to derive closed expressions for the energy and wave functions in Section 4.2. In fact, the value of the height of the potential for electrons should coincide with the discontinuity ΔE_c that appears at the interface in the conduction bands of AlGaAs and GaAs, which for the above system is of the order of some tenths of an eV. However, it is fairly easy to deduce that for electron energies in the quantum well not too close to the barrier ΔE_c (we take the energy as zero at the bottom of the well), the values obtained for the case of infinite wells do not differ too much of those obtained in the case of finite depth wells.

If we call V_0 in Figure 4.4 the height of the finite square well, it is evident that for states with energy $E < V_0$ we have *bound states*, i.e. the electrons are trapped inside the well of width a, while for $E > V_0$, we have continuous *propagation states*, in which the electrons are free to move from $z = -\infty$ to $z = +\infty$. Since this problem presents inversion symmetry around the centre of the well, this point is chosen as origin for the z-direction. In relation to the bound states, the wave functions inside the well should have the same shape as in the case of the infinite well, i.e. the solutions are symmetric or antisymmetric, and therefore should be sine or cosine functions, respectively. We also know from quantum mechanics that the solutions outside the well, which are obtained from the Schrödinger equation with a potential energy equal to V_0, are exponential decay functions. Therefore the solutions for the wave functions should be linear combinations of the functions:

$$\psi_n(z) = \begin{cases} D\exp(kz), & z < -a/2 \\ C\cos(kz), C\sin(kz), & -a/2 < z < a/2 \\ D\exp(-kz), & z > a/2 \end{cases} \tag{4.8}$$

where $k = \left(\dfrac{2m_e^* E}{\hbar^2}\right)^{1/2}$ inside the well $\tag{4.9a}$

and $k = \left(\dfrac{2m_e^* (V_o - E)}{\hbar^2}\right)^{1/2}$ outside the well $\tag{4.9b}$

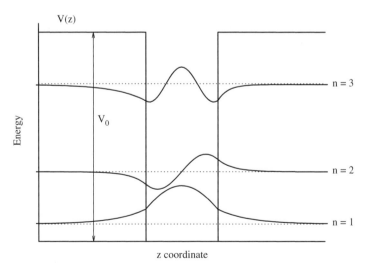

Figure 4.4. Finite potential square well. The first three energy levels and wave functions are shown.

Note that in Eqs (4.9) we have assumed the same value for the effective electron mass in the barrier and in the well. Usually, this is a good approximation because the barrier material is similar to the well material (e.g. GaAs and AlGaAs with a small Al mole fraction) and because the penetration of electron wave functions into the barriers is small for the lowest states. In order to proceed with the solution of the problem we should next realize that since $\psi(z)$ is a continuous function, therefore, the functions of Eq. (4.8) must be equal for $z = \pm a/2$, and the same should hold for their derivatives in the homogeneous effective mass approximation. From this fact a transcendental equation is derived which can be easily solved numerically [1]. From this solution, several important facts are derived. For instance, it can be shown that in a one-dimensional well there is always at least one bound state, independent of how small the value of V_0 might be. For the case of weakly bound states, the exponential decay constant k in Eq. (4.8) is small and therefore the wave function represented in Figure 4.4 penetrates deeply into the barrier region. Evidently, the opposite should be true for strongly bound states, in which the penetration in the energy forbidden region is very small.

4.4. PARABOLIC AND TRIANGULAR QUANTUM WELLS

4.4.1. Parabolic well

The case of the parabolic well is well known in solid state physics since the vibrations of the atoms in a crystal lattice, whose quantification gives rise to phonons, are described in a first approximation by harmonic oscillators. In addition, a magnetic field applied to a two-dimensional electron system gives rise to a parabolic potential, and the electrons oscillate at the so-called cyclotron frequency. Parabolic quantum well profiles can also be produced by the MBE growth technique. In this case, alternate layers of GaAs and $Al_xGa_{1-x}As$ of varying thickness are deposited, increasing the thickness of the AlGaAs layer quadratically with distance, while the thickness corresponding to the GaAs layer is proportionally reduced.

For the well-known potential energy of the one-dimensional harmonic oscillator

$$V(z) = \frac{1}{2}k^2 z^2 \tag{4.10}$$

with k a constant, we already know that the energy values are equidistant and given by

$$E_n = \left(n - \frac{1}{2}\right)\hbar\omega_0, \quad n = 1, 2, 3, \ldots \tag{4.11}$$

where ω_0 is the so-called natural angular frequency.

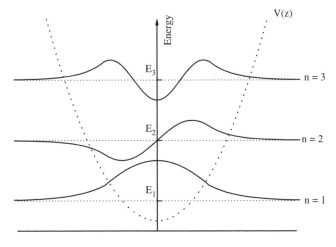

Figure 4.5. Harmonic oscillator potential well $V(z)$. The first three levels and corresponding wave functions are shown.

Figure 4.5 shows the case of a parabolic well potential and the wave functions of three electron bound states. The wave functions are mathematically expressed in terms of the Hermite polynomials. Note, as in the case of the square well, the symmetric or antisymmetric character of the wave functions and their exponential decay in the forbidden energy zone.

4.4.2. Triangular wells

The triangular potential well is one of the most common geometries, since the potential profile across quantum heterojunctions (Chapter 5), such as the well-known modulation-doped AlGaAs–GaAs heterojunction, is almost triangular in shape for electrons within GaAs. Of all heterostructures, this is probably the most investigated one and it will be considered in detail in Section 5.3.1. Another very important case, where an almost triangular-shaped well is formed, occurs at the semiconductor in a MOS structure (Section 5.2).

Figure 4.6 shows a triangular potential well, in which, for simplicity, it is assumed that the left barrier is infinite in energy and it increases linearly for $z > 0$:

$$V(z) = eFz, \quad \text{for } z > 0 \tag{4.12a}$$

$$V(z) = \infty, \quad \text{for } z \leq 0 \tag{4.12b}$$

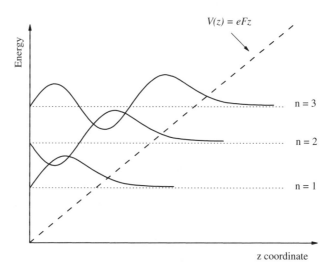

Figure 4.6. Triangular potential well $V(z)$. The first three levels and corresponding wave functions
are shown.

where e is a constant equal to the electron charge and F is a uniform electric field. As in the
other cases, the electron energies and states are found by solving the Schrödinger equation
subject to the boundary condition $\psi(z = 0) = 0$. In this case, the eigenvalues are given
in terms of the Airy functions. However, for small values of n, it can be demonstrated by
applying the WKB quantum-mechanical approximation, that [1]

$$E_n \approx \left[\frac{3}{2}\pi\left(n - \frac{1}{4}\right)\right]^{2/3}\left(\frac{e^2 F^2 \hbar^2}{2m}\right)^{1/3}, \quad n = 1, 2, \dots \tag{4.13}$$

Figure 4.6 shows the spacing between the energy levels, which get a little closer as n
increases, in contrast to the square well where the levels become further apart as n increases
(in the parabolic case they were equally spaced). In the same figure, the wave functions
are also represented. Observe that as n increases, the wave functions add one more half-
cycle. However at difference with the previous parabolic case, the wave functions are
neither symmetric or antisymmetric due to the asymmetry of the potential well.

4.5. QUANTUM WIRES

Having considered the two-dimensional electron gas in Section 4.2 it is easy to understand that in the one-dimensional electron gas, the electrons should be confined in two

directions, (x, y), and can freely propagate along the z-direction, usually perpendicular to the plane defined by x and y. The propagation is therefore somewhat analogous from a formal point of view to that of an electromagnetic wave guide.

Supposing that the confining potential is a function of $r = (x, y)$, i.e. $V = V(r)$, and following the separation of variables method to solve the Schrödinger equation, as we did in Section 4.2, we can look for wave functions of the form:

$$\psi(r) = e^{ik_z z} u(r) \tag{4.14}$$

with the following two-dimensional Schrödinger equation for the wave function $u(r)$:

$$\left[-\frac{\hbar^2}{2m_e} \left(\frac{\partial}{\partial x^2} + \frac{\partial}{\partial y^2} \right) + V(r) \right] u_{n_1 n_2}(r) = E_{n_1 n_2} u_{n_1 n_2}(r) \tag{4.15}$$

with $(n_1, n_2) = 1, 2, 3, \ldots$

The expression for the total energy of the electrons in the quantum wire should be of the form:

$$E_{n_1, n_2}(k_z) = E_{n_1, n_2} + \frac{\hbar^2 k_z^2}{2m_e^*} \tag{4.16}$$

where the last term represents the kinetic energy of the electron propagating along the z-direction.

As an example to get specific expressions for the energy, we consider now the simplest case of a two-dimensional rectangular potential of infinite depth and size a_x, a_y. That is:

$$\begin{aligned} V(x, y) &= 0, & 0 < x < a_x,\ 0 < y < a_y \\ V(x, y) &= \infty, & x \leq 0,\ x \geq a_x,\ y \leq 0,\ y \geq a_y \end{aligned} \tag{4.17}$$

Taking into account the results of Section 4.2, the energy should be given by Eq. (4.16) in which:

$$E_{n_1, n_2} = \frac{\hbar^2 \pi^2}{2m_e^*} \left(\frac{n_1^2}{a_x^2} + \frac{n_2^2}{a_y^2} \right), \qquad n_1, n_2 = 1, 2, 3, \ldots \tag{4.18}$$

Another case, relatively easy to solve in cylindrical coordinates, is that of a wire with circular cross section, in which case the solutions are given in terms of the Bessel functions. Therefore in the case of quantum wires, the energy levels corresponding to the transverse direction are specified by two quantum numbers, and each value E_{n_1, n_2} is now the bottom of a parabolic one-dimensional subband in k_z space. Observe also that as the electron moves in a narrower wire, the energy corresponding to the E_{n_1, n_2} levels increases.

Let us now calculate the density of states for the one-dimensional electron gas. The concentration of states in energy is related to that in wave number by the expression:

$$n_{1D}(E)\partial E = n_{1D}(E)\frac{dE}{dk}\partial k = 2n_{1D}(k)\partial k \qquad (4.19)$$

The factor 2 appears because the wave number could be either positive or negative corresponding to the two directions along the wire. The density of states in k-space per unit length is $1/2\pi$, as seen from considering the one-dimensional version of Figure 4.2. Substituting in Eq. (4.19) and taking into account that $E = \hbar^2 k^2/2m_e^*$, we obtain:

$$n_{1D}(E) = \frac{1}{\pi\hbar}\sqrt{2m_e^*/E} \qquad (4.20)$$

which diverges for $E = 0$

In terms of the group velocity v_g, given by Eq. (2.52):

$$n_{1D}(E) = \frac{2}{\pi\hbar v_g} \qquad (4.21)$$

One interesting result of this equation is that the current in a one-dimensional system is constant and proportional to the velocity and the density of states. The expression (4.21) for the DOS function in a quantum wire, will have important consequences as, for example, the quantized conductance which will be studied in Chapter 6.

The expression of the total DOS per unit length for a quantum wire can be expressed, from Eq. (4.20) as

$$n_{1D}(E) = \sum_{n_1,n_2} \frac{1}{\pi\hbar}\sqrt{\frac{2m_e^*}{E - E_{n_1,n_2}}} \qquad (4.22)$$

Figure 4.7 represents the DOS for a one-dimensional system, which is compared to the parabolic 3D case. Now the DOS diverges at the bottom of the subbands given by the energy values E_{n_1,n_2}. This result will have important consequences in the physical properties of quantum wires.

4.6. QUANTUM DOTS

Quantum dots are often nanocrystals with all three spatial dimensions in the nanometre range. Sometimes, as is the case of the II-VI materials, such as CdSe or CdS, the nanocrystals can be grown from liquid phase solutions at well-specified temperatures. Conversely, they can also be prepared by lithographic etching techniques from macroscopic materials.

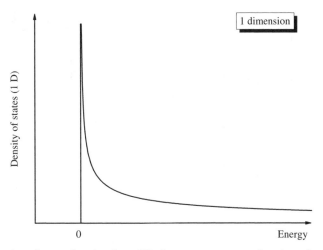

Figure 4.7. Density of states function for a 1D electron system, as a function of energy.

Although the word "dot" implies an infinitely small size, in practice dots might have a large number of atoms: 10^4–10^6, and still have their three dimensions in the nanometre region, so that the electron de Broglie wavelength is comparable to the size of the dot. In this case, the wave nature of the electron becomes important. Quantum dots are often referred to as *artificial atoms,* because, as we will see later, the spectrum of the energy levels resembles that of an atom. In addition, at least in theory the energy spectrum can be engineered depending on the size and shape of the dot. In analogy to atoms, we can also define an ionization energy, which accounts for the energy necessary to add or remove an electron from the dot. This energy is also called the *charging energy* of the dot, in an image similar to the concept of capacitance of a body, in which the addition or subtraction of electric charge is specified by the Coulomb interaction. Therefore the atom-like properties of the quantum dots are often studied via the electrical characteristics. From this point of view, it is very important to remark that even the introduction or removal of one single electron in quantum dots, in contrast to the case of 2D or 1D systems, produces dominant changes in the electrical characteristics, mainly manifested in large conductance oscillations and in the Coulomb blockade effect (Chapter 6).

Let us now study the energy spectrum of quantum dots. The simplest case would be that of a confining potential that is zero inside a box of dimensions a_x, a_y, and a_z and infinite outside the box. Evidently the solution to this well-known problem are standing waves for the electron wave function and the energy levels are given by

$$E_{n_1,n_2,n_3} = \frac{\hbar^2 \pi^2}{2m_e^*} \left(\frac{n_1^2}{a_x^2} + \frac{n_2^2}{a_y^2} + \frac{n_3^2}{a_z^2} \right), \qquad n_1, n_2, n_3 = 1, 2, 3, \ldots \qquad (4.23)$$

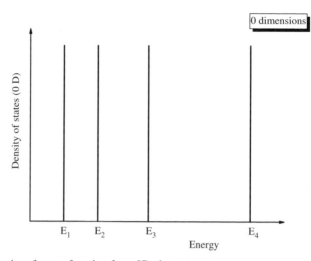

Figure 4.8. Density of states function for a 0D electron system.

In contrast to the 2D and 1D cases, now the energy is completely quantized, and, as in the case of atoms, there is no free electron propagation. However, the levels are frequently degenerate, for instance, if two or three of the dimensions of the box are equal.

The case of a spherical dot in which the potential is zero inside the sphere and infinite outside can also be exactly solved, and the solutions are expressed in terms of the spherical functions. This problem resembles that of a spherically symmetric atom and the energy depends on two quantum numbers, the principal quantum number n, arising from the one-dimensional radial equation, and the angular momentum quantum number l.

Since in the case of quantum dots the electrons are totally confined, the energy spectrum is totally discrete and the DOS function is formed by a set of peaks in theory with no width and with infinite height (Figure 4.8). Evidently, in practice, the peaks should have a finite width, as a consequence, for instance, of the interaction of electrons with lattice phonons and impurities.

4.7. STRAINED LAYERS

In general, the quality of an interface between two materials depends greatly on the relative size of the lattice constants. If the lattice constants are very similar, as in the case of the $Al_xGa_{1-x}As$–GaAs heterojunctions (see Figure 4.9), for which the lattice constant varies less than 0.2% for the whole range of x, and the thermal expansion coefficients are similar, no stresses are introduced at the interface. However, in other cases, heterojunctions with differences in lattice constants up to 6% are fabricated (for instance, $In_xGa_{1-x}As$–GaAs).

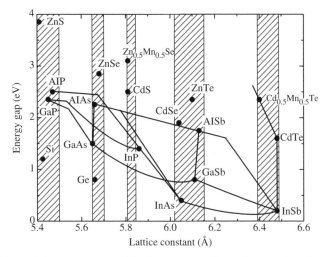

Figure 4.9. Values of the energy gap and lattice constants for various semiconductors. The shaded regions refer to typical families of semiconductors which form several types of heterostructures.

In this case strong stresses appear at the interface and only very thin films of a few monolayers can be grown on a given substrate. Nevertheless these strained layers show new effects that are exploited in various optoelectronic applications, especially in quantum well lasers and electro-optic modulators (Chapter 10). As a consequence, a new field of strained layer epitaxy has been developed. Usually during the growth of the layer over a substrate the lattice constant of the epilayer changes, and accommodates to the size of the lattice constant of the substrate, a situation which is called pseudomorphic growth. These stresses have very important effects such as the removal of the heavy and light hole degeneracy of the valence band (see Section 4.8), changes in the bandgaps, etc. since the structure of bands in solids is very sensitive to both changes in size as well as in the symmetry of the unit cell.

Suppose that a layer of lattice constant a_L is grown on a substrate of lattice constant a_S. The strain ε of the layer is defined as

$$\varepsilon = \frac{a_L - a_S}{a_S} \tag{4.24}$$

In Figure 4.10 we show the case of $In_xGa_{1-x}As$ grown on GaAs for which $a_L > a_S$. In Figure 4.10(a) the situation of the separate layer and substrate is depicted. If the epilayer is not too thick, the atoms of the layer match those of the substrate. Therefore, the layer is subjected to a compressive stress in the plane of the interface and the interplanar

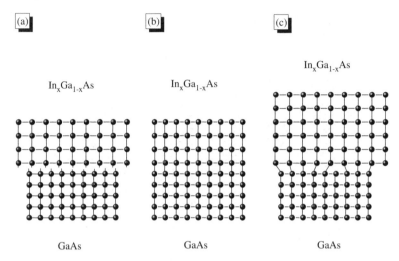

Figure 4.10. Growth of a layer of lattice constant a_L on a substrate of constant a_S for $a_L > a_S$. (a) The layer is not yet deposited over the substrate; (b) the layer is subjected to a compressive stress; (c) when the layer is thicker than the critical thickness, dislocations are formed at the interface.

vertical (or on the growth direction) distance increases (Figure 4.10(b)). As a consequence of the distortion of the InGaAs lattice, elastic energy is stored in the system, which increases with layer thickness. Therefore, if the thickness of the film is larger than the so-called *critical thickness* h_c, the system relaxes and the appearance of dislocations in the plane of the interface is energetically favourable (Figure 4.10(c)). Although it is possible to grow layers without dislocations with a thickness greater than h_c, evidently they are in a metastable state. Therefore the layers are usually grown up to a thickness below h_c. In addition to the $In_xGa_{1-x}As$ layers grown on GaAs, another system that has been the subject of many investigations is the Ge_xSi_{1-x}/Si system because of its applications in microelectronics.

4.8.   EFFECT OF STRAIN ON VALENCE BANDS

In order to study the electronic and optoelectronic properties of the most significant semi-conductors for solid state devices, for instance, silicon and III-V compounds, it is essential to have a good knowledge of the shape of the conduction and valence bands in k-space. For electrons in a crystal, the band states arise from the outermost states of the electrons in the atoms. In the case of the conduction band, the behaviour of electrons can often be described by band states, which in the case of direct gap materials, are purely s-type. Using the notation of atomic physics these are called 1s wave functions for the cell periodic part of the Bloch functions. However the band states of the indirect gap semiconductors are a mixture of s and p states.

The case of the valence band is more complicated because there are three branches, which are very close, around $k = 0$, and arise from the atomic p-states. This happens for many important semiconductors like Si, Ge, and GaAs, which show degenerate hole bands for $k = 0$, i.e. close to the maximum of the valence bands. In Figure 4.11, we can observe that one of the bands corresponds to the light holes, whose curvature is larger, and the other corresponds to the heavy holes and presents a smaller curvature. There is also a third band, called *split-off band*, which appears below the above ones, and is a consequence of the relativistic effects corresponding to the spin–orbit coupling. However, very often the splitting is quite large and this band can be ignored, since the holes would not fill the corresponding levels. Near the point Γ, the heavy- and light-hole bands depend closely in a parabolic fashion on k and therefore can be described by the so-called effective masses for the heavy holes (m_{hh}^*) and light holes (m_{lh}^*), respectively.

In order to discuss the E vs k dependence in the valence band, we can use the Luttinger–Kohn formulation, based on the $\boldsymbol{k} \cdot \boldsymbol{p}$ method of band structure calculations. The eigenstates in this formulation are the angular momentum p-states. Following the calculations of the *Luttinger–Kohn method*, it is found that the energy depends on k approximately in the following simple manner:

$$E(k) = E_v - \frac{h^2}{2m_0} \left[Ak^2 \pm \sqrt{\left(Bk^2\right)^2 + C^2 \left(k_x^2 k_y^2 + k_y^2 k_z^2 + k_z^2 k_x^2\right)} \right] \tag{4.25}$$

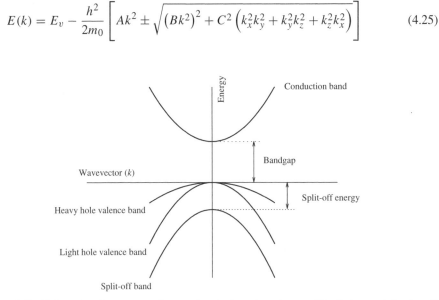

Figure 4.11. Valence band structure of several typical semiconductors (Si, GaAs, etc.) for $k = 0$. The heavy hole, light hole, and split-off bands are shown.

where the upper sign corresponds to the heavy holes and the lower sign to the light holes, respectively. In the above equation the dimensionless A, B, and C parameters are usually expressed in terms of the Luttinger–Kohn parameters γ_1, γ_2, and γ_3:

$$A = \gamma_1, \quad B = 2\gamma_2, \quad C = 12\left(\gamma_3^2 - \gamma_2^2\right) \tag{4.26}$$

In the case of GaAs we have the following values for these parameters: $\gamma_1 = 6.85$, $\gamma_2 = 2.1$, and $\gamma_3 = 2.9$.

The effect of strain on the band structure is very important especially in the hole valence bands, as we should expect from the changes in the layer lattice constant and crystal symmetry. As a consequence we can introduce modifications of the semiconductor bandgaps and the lifting of the valence band degeneracies at Γ. The shifts in energy of the electron and hole bands, as a consequence of the strain, can be calculated using the usual methods of energy band calculations.

The most significant effects in III-V compounds are manifested in the changes of the gaps and the structure of the valence bands. For layers grown in the z-direction or (0 0 1) axis, the band structure of an unstrained layer, one under compression and one under tension in the interface plane is represented in Figures 4.12(a)–(c), respectively [2]. As previously mentioned, the first observation is the removal of the heavy and light hole degeneracy. In the case of compression, the top valence band (heavy hole band) corresponds to the highest value of the effective mass m_z (HH) but the in-plane mass

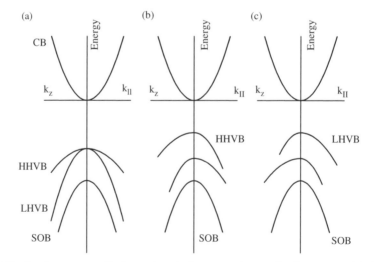

Figure 4.12. Band structures of a layer deposited over a substrate for three cases: (a) the layer is unstrained; (b) the layer is under compression; (c) the layer is under tension. After [2].

is somewhat lighter (Figure 4.12(b)). In the case of tension (Figure 4.12(c)), in which the lattice constant of the layer material in the bulk is smaller than that of the substrate we find the opposite ordering of the bands, i.e. the light hole band is above the heavy hole. Therefore the two bands will cross each other at some value of $k_{||}$, although such values are not usually reached ($k_{||}$ is any vector included in the plane k_x, k_y). As it is observed, strain induces band splittings of up to about 0.05 eV, in addition to anisotropy. Strained structures find applications in the change of the bandgaps in heterojunctions such as Si/SiGe, GaP/GaAsP, and in the fabrication of blue-green lasers based on ZnCdSe/ZnSe heterostructures.

4.9. BAND STRUCTURE IN QUANTUM WELLS

In order to interpret correctly the optical absorption experiments in quantum wells, we need to know the band structure. Figure 8.2 of Chapter 8 shows the absorption spectra of a 40 period multiple quantum well (MQW) GaAs–AlGaAs, in which the barriers have a width of 7.6 nm [3]. Observe that the spectrum follows in general the steps of the DOS curve in 2D semiconductors (Section 4.2). At the edge of each step there is a sharp maximum that, as will be shown in the next section, is attributed to excitonic effects. It can also be observed in Figure 8.2 that at the edge of the first transition for electrons between the conduction and valence bands for $n = 1$, there is a peak at 1.59 eV which corresponds to the heavy hole (HH) valence band and one at 1.61 eV for the light holes (LH), which is below the heavy ones (Section 4.8).

The reason for observing the above splitting corresponding to holes is that the one-dimensional potential due to the quantum well breaks the cubic symmetry of the crystal, and consequently lifts the degeneracy of the hole band in GaAs, in a similar manner as did strain in the previous section. Detailed calculations, too long to be included in this text, show that the presence of the well potential causes the LH states to move downwards in energy more than the HH (Figure 4.13). It is interesting to know that if the calculations do not take into account very small terms in the expansion of the perturbed Hamiltonian, the hole bands cross each other because, then, the heavy hole band moves faster downwards. The resulting crossing of the two bands, produced as a consequence of their different curvature, would cause the phenomenon known as mass reversal. If this does not happen, it is because, when very detailed calculations are performed, it can be shown that the crossing effect is removed, appearing instead as an effect known as *anticrossing*.

We can therefore appreciate that the band structure in quantum wells, especially those corresponding to holes, are fairly complicated and most results can only be numerical. An additional complication comes from the fact that the square wells have barriers of finite height (Section 4.3). Figure 4.14 shows the calculated band structure for MQW of the type AlGaAs–GaAs [4]. Note that some of the bands have a shape far from the ideal

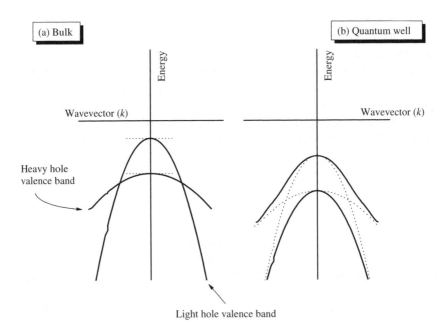

Figure 4.13. (a) Valence bands of GaAs bulk crystal; (b) position of the valence bands in a GaAs in a quantum well.

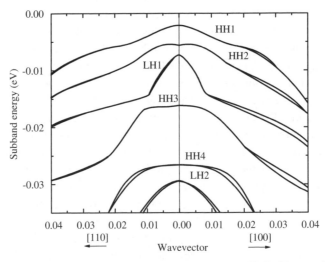

Figure 4.14. Structure of the valence bands in a GaAs–AlGaAs MQW. Observe that some bands (HH2) show negative hole masses close to $k = 0$ and that some of the bands (HH3) have a shape far from parabolic. After [4].

parabolic one and that the sign of the curvature changes, implying the existence of hole states with a negative effective mass.

As we have seen in this section, quantum confinement can cause a change of the energy levels corresponding to the HH and LH bands. In addition, we can change the energy of these bands by applying tensions or compressions (Section 4.8). In some cases, we can move the LH band above the HH, or even cause degeneracy. As a consequence of the degeneracy, a very high concentration of hole states is produced, a phenomenon which is exploited in optoelectronic modulators.

4.10. EXCITONIC EFFECTS IN QUANTUM WELLS

In the case of confined systems for electrons and holes, such as quantum wells, wires, and dots, the excitonic effects are much more important than in bulk solids. In effect, the *binding energy* of the electron–hole system forming an exciton is much higher in quantum confined systems than in the case of solids, and, therefore, the excitonic transitions can be observed even at temperatures close to room temperature, as opposed to the bulk case for which low temperatures are needed. This makes the role played by excitons in many modern optoelectronic devices very important.

Qualitatively, it is easy to understand the reason by which the binding energies of excitons E_B in quantum confined systems are much higher than in the bulk. For instance, in the case of bulk GaAs the binding energy of excitons is only $E_B = 4.2$ meV and the Bohr radius has a fairly large value of about $a_B = 150$ Å. In Figure 4.15 the exciton is represented in two situations, the one in (a), in which the Bohr radius of the exciton is much smaller than the quantum well width, and the one in (b), for which the width of the well is smaller than a_B. In this case, the separation between the electron and the hole is limited by the width of the well and therefore the exciton becomes squeezed, thus increasing the Coulomb attractive force. For a two-dimensional hydrogenic atom, a simple calculation shows that E_B is about four times larger in the 2D case than for a 3D solid. Another consequence of their high value of E_B is that excitons can survive very high electric fields in quantum wells which will find many applications in electro-optic modulators (Section 10.8). In addition, the oscillator strength for excitonic transitions appears in a narrow energy range, thus increasing the intensity of the excitonic transitions.

More realistic calculations based on numerical methods, rather than on the two-dimensional hydrogen atom, allow for the calculation of the exciton binding energy as a function of the quantum well width. For the $Al_xGa_{1-x}As$–GaAs quantum well, the results are shown in Figure 4.16 as a function of well width [5]. If this is much larger than the bulk exciton Bohr radius, the exciton has a binding energy similar to the bulk. As the width of the well decreases, the exciton squeezes inside the well, and the Coulomb interaction increases. For a well width of 3–4 nm, the binding energy reaches a maximum

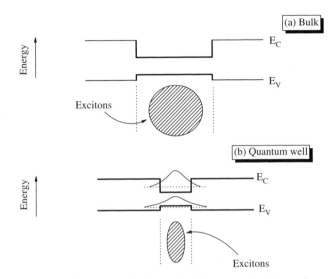

Figure 4.15. (a) Exciton orbits in a very bulk crystal; (b) the spherical form of the exciton becomes elongated when the width of the QW is smaller than the exciton radius.

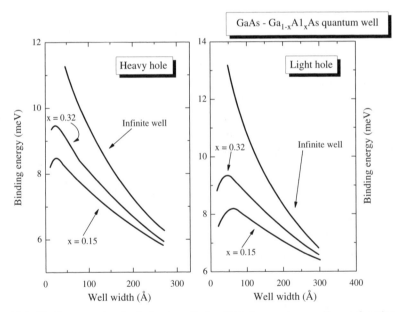

Figure 4.16. Binding energies for holes in a GaAs–AlGaAs quantum well as a function of well width. After [5].

value, which is close to three times that in the bulk. When the well of finite height becomes too thin, the exciton confinement energy increases and the electron (and to a lesser extent the hole) wave functions leak into the barriers due to tunnelling. Then the effective localization of the exciton in the well decreases and its binding energy also decreases towards the bulk value of the barrier material ($Al_xGa_{1-x}As$). It can be shown theoretically that in the case of quantum wires (1D confinement) and quantum dots (0D confinement), the effect on the exciton binding energy is even more important than for quantum wells (2D confinement). However, fabrication technologies for the sub-2D systems are not as developed as for the 2D ones, and the effect of higher binding energies for excitons are not usually exploited. One exception might be the case of quantum dot lasers, which will be treated in Section 10.6.

REFERENCES

[1] Davies, J.H. (1998) *The Physics of Low-Dimensional Semiconductors* (Cambridge University Press, Cambridge).
[2] Reilly, E.P.O. (1989) *Semiconductor Science and Technology*, **4**, 121.
[3] Fox, A.M. (1996) *Contemporary Physics*, **37**, 111.
[4] Chang, Y.C. & Shulman. J.N. (1985) *Phys. Rev. B*, **31**, 2069.
[5] Greene, R.L., Bajaj, K.K. & Phelps, D.E. (1984) *Phys. Rev. B*, **29**, 1807.

FURTHER READING

The pioneering text of Jaros introduces in a simple way some of the concepts of nanostructures. The book of Weisbuch and Vinter and that of Kelly are excellent and treat at an intermediate level most of the fundamentals and applications of low-dimensional semiconductors. The book of Davies is very enlightening in treating the fundamental aspects of semiconductor nanostructures. The book of Mitin et al. is also excellent in the treatment of fundamentals and applications of semiconductor quantum heterostructures (however, it does not treat the properties under magnetic fields or the quantum Hall effect). The level of the book of Ferry and Goodnick is more advanced; although the emphasis is on transport properties, it also includes introductory chapters on fundamentals.

Davies, J.H. (1998) *The Physics of Low-Dimensional Semiconductors* (Cambridge University Press, Cambridge).
Ferry, D.K. & Goodnick, S.M. (1997) *Transport in Nanostructures* (Cambridge University Press, Cambridge).
Jaros, M. (1989) *Physics and Applications of Semiconductor Microstructures* (Clarendon Press, Oxford).

Kelly, M.J. (1995) *Low-Dimensional Semiconductors* (Clarendon Press, Oxford).
Mitin, V.V., Kochelap, V.A. & Stroscio, M.A. (1999) *Quantum Heterostructures* (Cambridge University Press, Cambridge).
Weisbuch, C. & Vinter, B. (1991) *Quantum Semiconductor Structures* (Academic Press, Boston).

PROBLEMS

1. **Electronic levels in a quantum well**. Suppose a quantum well with a width a and a conduction band discontinuity large enough so that the model of a box with infinite wells can be applied. Explain how the effective mass of the semiconductor of the well affects the values of the energy levels in the well. Do it for InSb ($m^* = 0.014m_0$), GaAs ($m^* = 0.067m_0$), and GaN ($m^* = 0.17m_0$).

2. **Band gap in quantum well systems**. Suppose a quantum well of the type $Al_xGa_{1-x}As/GaAs/Al_xGa_{1-x}As$ with $x \approx 0.3$. Calculate the bandgap if the width a of the well is 16 nm. (b) Comment whether the model of infinite wells is appropriate for this value of a and also in the case of very small values of a, like, for instance, 2 nm.

3. **Fermi wavelength**. The Fermi wavelength λ_F is defined as the wavelength of electrons with energy equal to the Fermi energy E_F. Calculate the values of λ_F for electrons in the case of a quantum well of the type AlGaAs/GaAs/AlGaAs for $n_{2D} = 10^{16} m^{-2}$. Compare the result obtained with that of λ_F for a typical metal.

4. **2D density of states function**. Show that the 2D density of states function per unit area n_{2D} (Figure 4.3) can be expressed as

$$n_{2D} = \frac{m^*}{\pi \hbar^2} \sum_n \Theta(E - E_n)$$

    where $\Theta(z)$ is the step function defined as $\Theta(z) = 1$ for $z > 0$, $\Theta(z) = 0$ for $z < 0$, and the sum is over the subbands.

5. **Concentration of electrons in a quantum well**. From the expression of the density of states function and the Fermi–Dirac distribution function, show that the concentration n_{2D} of electrons in a 2D system is given by:

$$n_{2D} = \frac{kTm^*}{\pi \hbar^2} \sum_n ln(1 + e^{(E_F-E_n)/kT})$$

    where E_F is the Fermi energy and E_n are the values of the energy levels in the quantum well.

6. **Extreme quantum limit in a 2D system**. From the expression for the concentration n_{2D} of electrons of the previous exercise, show that in the limit when only the first subband ($n = 1$) is occupied, then

$$n_{2D} = \frac{m^*}{\pi \hbar^2} E_F$$

where E_F is the Fermi energy. Additionally show that the Fermi wave vector k_F is given by

$$k_F = (2\pi n_{2D})^{1/2}$$

7. **Density of states in k-space**. Show that the density of states in k-space $g(k)$ depends on k in the form: $g(k) \propto k^{n-1}$ where n is the number of dimensions of the system. *Hint*: assume that the values of the components of $\vec{k}$ are obtained by applying periodic boundary conditions, as it was done in Section 2.3 for $n = 3$.

8. **Square potential well of finite height**. Suppose a finite square well potential with potential barriers of height V_0 and width a for which the wave functions are of the form of Eqs (4.8) and (4.9). Show that, in order to have bound states in the well, the following conditions have to be satisfied

$$\tan(k_w a/2) = \frac{m_w^* k_b}{m_b^* k_w}$$

$$\tan(k_w a/2) = -\frac{m_b^* k_w}{m_w^* k_b}$$

where the subindexes w, b refer to the well material and barrier material, respectively. *Hint*: assume as boundary conditions that both the wave function and the particle flux $(1/m^*) \, d\psi/dz$ must be continuous.

9. **Realistic AlGaAs/GaAs/AlGaAs quantum well**. Based on the results of the previous problem find the electron ground state energy for a AlGaAs/GaAs/AlGaAs quantum well of width $a = 12$ nm and height of the potential well $V_0 = 0.36$ V. Use for the effective masses $m_w^* = 0,067 m_0$ and $m_b^* = 0.095 m_0$. *Hint*: in the results of the previous problem, write the values of the wave vectors in terms of energy using Eq. (4.9). Since there in no analytic solution for E, it is recommended that the problem be solved graphically.

10. **Structures of reduced dimensionality**. Suppose a semiconductor quantum well of width a, a quantum wire of square cross section ($a \times a$) and a quantum dot of volume ($a \times a \times a$). In the case of GaAs, and for $a = 6$ nm, calculate the energy E_{el} of the electron ground level for each structure. Qualitatively comment why the value of E_{el} gets larger when the confinement of the structure increases.

11. **Fermi energy in low-dimensional electron systems**. Show that for 3D, 2D, and 1D electron systems for which the energy E depends on the wave vector k and the effective mass m^* in the form $E = \hbar^2 k^2 / 2m^*$, the Fermi energy E_F can be expressed as

$$3D: \quad E_F \;=\; \frac{\hbar^2}{2m^*}\, (3\pi^2 n)^{2/3}$$

$$2D: \quad E_F \;=\; \frac{\hbar^2}{2m^*}\, 2\pi n$$

$$1D: \quad E_F \;=\; \frac{\hbar^2}{2m^*}\, \left(\frac{\pi}{2}\, n\right)^2$$

where n corresponds, in each case, to the 3D, 2D, and 1D electron concentration.

Chapter 5

Semiconductor Quantum Nanostructures and Superlattices

Chapter 5

Semiconductor Quantum Nanostructures and Superlattices

5.1. INTRODUCTION

The previous chapter described the physical behaviour of electrons in low-dimensional semiconductors. In this chapter we will study a series of nanostructures or devices formed by one or several heterojunctions, which constitute the basis of modern microelectronic and optoelectronic devices. These devices can be constructed from a few basic structures, among them heterojunctions, or junctions between two semiconductors of different gaps, and metal-oxide-semiconductor (MOS) structures. These two kinds of nanostructures provide electrons with a potential well for electrons of nanometric size at the junction. Since the MOS structure is the basic block of the most important device in microelectronics, the metal-oxide-semiconductor field-effect-transistor (MOSFET), it is natural that researchers have made use of them for the study of the behaviour of the electrons in nanometric potential wells. In fact, the discovery of the quantum Hall effect (QHE) by K. von Klitzing in 1980 was based on the study of the transport properties of electrons in the channel of a MOSFET under the influence of simultaneous electric and magnetic fields [1]. However, these new quantum effects (QHE, Aharonov–Bohm effect and Shubnikov–de Haas oscillations, Chapter 7) are better observed in III-V heterojunctions, since the electron effective mass is much lower in materials such as GaAs than in Si.

In this chapter we shall study first the behaviour of electrons confined in 2D wells at the semiconductor-oxide interface in MOSFET transistors. Next, we shall proceed with the III-V modulation-doped heterojunction used in high-frequency transistors. The strained SiGe heterojunction will also be considered because of its interest. Next, we will focus on the modulation-doped, square potential, quantum wells. This simple building block is used as a single unit or, more often, as a multiple quantum well structure in devices. When the thickness of the barriers separating the wells is small, tunnelling of electrons between neighbouring wells takes place and the resulting device, called a superlattice (SL), displays a band diagram similar to that of electron bands in crystals. However, the allowed energy bands and gaps correspond to much smaller energy intervals, since the SL has a spatial periodicity (equal to the sum of well and barrier thicknesses) which is much larger than the lattice constant. The band structure of a SL can be engineered by a proper choice of the well and barrier widths. In this sense, SLs can be considered as artificial solids since their electron energy band structure is similar to the ones in crystals, but they do not exist in nature. The idea of developing SLs was first proposed by Esaki and Tsu in 1970 and implemented a few years later.

5.2. MOSFET STRUCTURES

The main contribution to present technology in general, and microelectronics in particular, is probably the metal-oxide-semiconductor field-effect-transistor (MOSFET). This device is the basic unit of present ultra-large-scale-integration (ULSI) microelectronics industry. It is estimated that MOSFET-based electronic devices now constitute close to 90% of the semiconductor device market. The MOSFET is formed by a MOS structure and two p–n$^+$ junctions in which the n material is heavily doped (Figure 5.1(a)), which act as the source and drain of the FET. The gate of the transistor is formed by the MOS structure. The semiconductor is usually p-type silicon over which a thin oxide layer (gate oxide) is grown by thermal oxidation.

Figure 5.1(b) shows the band diagram of the MOS structure for a p-type silicon semiconductor under a fairly strong positive bias. When a positive potential is applied to the gate, the electrons coming from the n$^+$ regions and some from the bulk p-silicon are accumulated at the Si–SiO$_2$ interface. These electrons form the so-called *inversion layer or channel* and are located in an almost triangular-shaped potential well of nanometric dimensions. The shape of the well is due to the space charge of ionized acceptors in the p-type silicon, whose corresponding holes are repelled by the electric field across the dielectric oxide produced by the positive gate potential. If a positive potential is applied between drain and source, the electrons in the channel will create a current. The current can be modulated by changes at the potential gate, since the amount of electrons in the

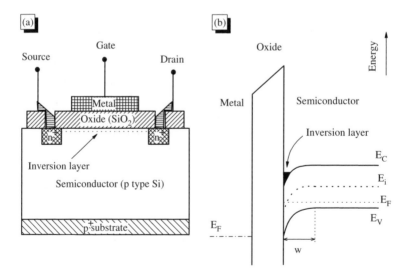

Figure 5.1. (a) Structure of a MOS transistor; (b) band diagram of a MOS structure with a positive strong bias applied to the gate.

inversion layer depends on the magnitude of the electric field across the insulator. The name "Field Effect Transistor" is due to the control effect of the electric field applied to the gate.

The existence of a mature MOSFET technology and a strong microelectronic industry is largely due to the excellent properties of the Si–SiO$_2$ interface and the possibilities of reducing the structure of the MOSFET to sizes below 100 nm in modern integrated circuits. In 1957, Schrieffer had already pointed out the possible quantification of the electrons in nanometric quantum wells, but it was not until a decade later that quantification was really observed in a silicon surface. In order to observe quantum effects at the Si–SiO$_2$ interface, several conditions must be met: (a) the SiO$_2$ insulator, of amorphous nature, should have neither high concentration of impurities (Na$^+$ ions) nor of trapped charge; (b) the smoothness of the Si–SiO$_2$ interface should be controlled at the atomic size level, since a rough surface on top of the inversion channel would greatly decrease the electron mobility in the inversion layer.

In order to study the behaviour of electrons in the potential well, it should be recognized that the electron inversion layer can be considered a 2D system of electrons immersed in a triangular-shaped quantum well (Section 4.4.2), located in the semiconductor, close to the interface with the oxide. In the MOS structure, although the electrons are confined along the perpendicular direction, they are practically free to move in the plane of the interface. Therefore, according to Section 4.2, the quantized values for the energy of confinement should be given by

$$E = E_n + \frac{\hbar^2}{2m_x^*}k_x^2 + \frac{\hbar^2}{2m_y^*}k_y^2 \tag{5.1}$$

where E_n corresponds to the quantized energy for the triangular well (Eq. (4.13)).

$$E_n \approx \left[\frac{3}{2}\pi\left(n - \frac{1}{4}\right)\right]^{2/3}\left(\frac{e^2F^2\hbar^2}{2m_z^*}\right)^{1/3}, \quad n = 1, 2, \ldots \tag{5.2}$$

Evidently Eq. (5.1) represents parabolas as in Figure 4.1(c) in reciprocal space, the bottoms having values given by E_n. Similarly, the density of states (DOS) function corresponds to the 2D case and is given by

$$g(E) = g_v \frac{m_T^*}{\pi \hbar^2} \tag{5.3}$$

where we have added the factor g_v which takes into account the conduction band valley degeneracy. This degeneracy arises from the fact that constant energy surfaces of the silicon conduction band are formed by six ellipsoids in the <001> direction of momentum space

(Figure 3.3). The long axis of the ellipsoid corresponds to the longitudinal effective mass $m_L^* = 0.91m_0$ and the two equal short axes to the transversal effective mass $m_T^* = 0.19m_0$, which is the one that appears in Eq. (5.3). Therefore, there are two conduction band minima corresponding to the heavy effective mass and four to the light one. As a consequence, after solving the Schrödinger equation, neglecting coupling of the electrons in the various conduction band minima, one should expect two different subband values, or subband ladders for the Si <001>. First consider the electrons in valleys perpendicular to the interfaces. The effective mass which enters Eq. (5.2) is m_L^*. Also, $m_x^* = m_y^* = m_T^*$ in Eq. (5.1) and $g_v = 2$. This results in subbands of lower energy (higher value of effective mass). Evidently the second subband ladder is originated when the parallel valleys are considered. In this case, m_T is the effective mass for the expression of E_n in Eq. (5.2) and $g_v = 4$.

Figure 5.2 shows the results obtained for the energies for the case of a Si-MOSFET with the insulating oxide grown over a <001> silicon surface [2]. The results are obtained

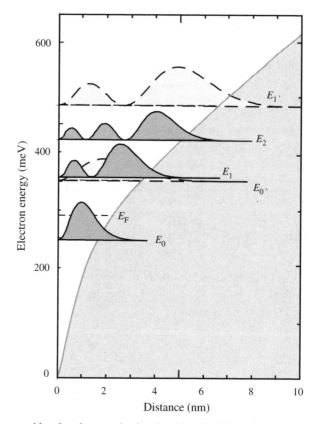

Figure 5.2. Energy subband series, conduction band profile $V(z)$ and squared wave functions for the potential well in a MOS structure for Si <001>. After [2].

by the self-consistent numerical solution of Schrödinger and Poisson equations using the parameters $m_L^* = 0.916m_0$ and $m_T^* = 0.190m_0$. The subband series E_0, E_1, E_2 correspond to m_L^* and $E_{0'}$, $E_{1'}$, $E_{2'}$, to m_T^*. The solid curve represents the conduction band profile $V(z)$. The squared wave functions for the two subband series are also represented. As seen in Chapter 7, there is now enough experimental evidence (quantum Hall effect and Shubnikov–de Haas effect) which shows that the 2D electrons in the MOSFET inversion layer are quantified.

5.3. HETEROJUNCTIONS

5.3.1. *Modulation-doped heterojunctions*

Heterojunctions, or interfaces between two semiconductors of different gaps, are one of the most versatile building blocks of electronic devices, especially those based on III-V compounds. Probably the most studied heterostructure is the one formed by n-type $Al_xGa_{1-x}As$ and almost intrinsic or lightly doped p-type GaAs. In a similar fashion to the case of the MOS (Section 5.2), an inversion layer of electrons is formed in the GaAs close to the GaAs–AlGaAs interface. Therefore, the physics of this electronic system should in principle be very similar to the case of the MOS structure studied in the previous section. Devices based on AlGaAs structures can be used to much higher frequencies than silicon devices due to the high mobility of electrons in GaAs. Since oxides and insulators deposited over GaAs do not present an interface of sufficient quality, the most important device applications are based on a Schottky structure of the type metal–AlGaAs–GaAs (Figure 5.3(a)). In this section we will focus on the properties and band diagram of the $Al_xGa_{1-x}As$–GaAs heterojunction shown in Figure 5.3(b).

Let us first consider, from a qualitative point of view, how an electron well of nanometric size is formed at the AlGaAs–GaAs interface. Suppose, as in Figure 5.4, that we have an AlGaAs–GaAs heterojunction, where the left material is gallium arsenide doped with aluminium and the right one is near-intrinsic GaAs. This structure is called a modulation-doped heterojunction and the method to produce it is known as modulation doping. First consider the hypothetical situation of Figure 5.4(a), before the two semiconductors enter in contact. In the figure, for simplicity, we only draw the bottom of the conduction bands and the Fermi level, which in the case of n-type AlGaAs is close to the conduction band, and for lightly p-doped GaAs is located close to the middle of the gap (Section 3.4). Evidently the bands are flat because the materials are electrically neutral and have uniform doping. The barrier between them in the conduction band, ΔE_c, can be approximately found following Anderson's rule. According to this rule, when we join two materials, the vacuum levels should line up. If χ_A and χ_B are the electron affinities of the AlGaAs and GaAs, respectively, we should have $\Delta E_c \equiv \chi_A - \chi_B$, since the electron affinity of a semiconductor is defined as the energy required for an electron located at the

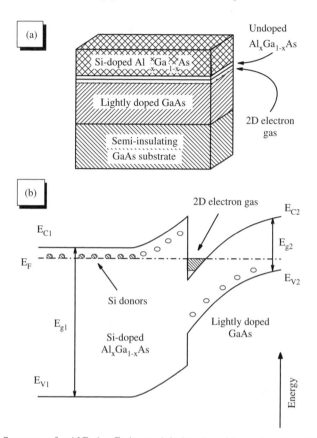

Figure 5.3. (a) Structure of a AlGaAs–GaAs modulation-doped heterojunction; (b) corresponding band diagram.

bottom E_c of the conduction band, to get out of the solid, i.e. $\chi = E_{vac} - E_c$. According to this rule, one gets a value of ΔE_c of 0.35 eV for a doping x in $Al_x Ga_{1-x}As$ around 0.3.

When both materials, AlGaAs and GaAs, enter in contact, some of the electrons from the donors of the n-material will cross the interface reaching the undoped GaAs. Therefore, as in the p–n junction, an internal electric field will be created and directed from the non-neutralized donors in the AlGaAs to the additional electronic charges in the GaAs. This field is the one that causes the band bending shown in Figure 5.3(b). At equilibrium, the two Fermi levels line up, the bands are bent like in the case of the p–n junction, with the only difference that the barrier ΔE_c is created. Note also that far from the interface, the bottom E_c of the conduction bands is flat and at the same distance from the Fermi level E_F as in the case of Figure 5.4(a). Therefore, it is relatively easy to sketch the band diagram of Figure 5.4(b). As it can be appreciated, a quantum well for the electrons has

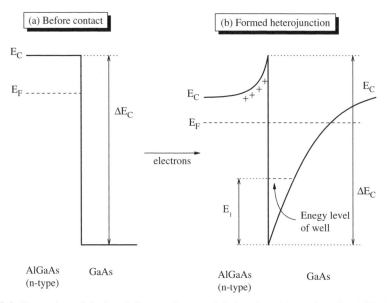

Figure 5.4. Formation of the band diagram in a modulation-doped heterojunction AlGaAs–GaAs when both semiconductors enter in contact.

been formed which is limited by a potential well of height ΔE_c in the left and a potential energy curve at the right, within the GaAs.

The quantum well for the electrons produced at the AlGaAs–GaAs interface has a shape close to a triangle as in the case of the MOS structure. Therefore, if we call z the direction perpendicular to the interface, the electrons forming the 2D inversion layer are free to move along the (x, y) plane, but their energy for the motion along z is quantized as in a potential well. The most important aspect of this heterojunction is that the charge carriers are located in a region (mainly in the GaAs), spatially separated from the AlGaAs semiconductor which originates the free electrons. The electrons in the well should have very high mobility for their motion along the (x, y) plane, since they move within the GaAs which is free of dopant impurities and it is well known that impurity scattering is one of the main factors which limit carrier mobility, especially at low temperatures. Evidently, the electron mobilities are also much higher than in the case of the MOS structure studied in the previous section.

Although the above considerations are mainly qualitative, they allow us to describe the main properties of AlGaAs–GaAs heterojunctions. For a rigorous treatment, one has to numerically solve Poisson's equation for the potential and Schrödinger's equation for the electron wave functions, following a self-consistent method. Although somewhat complicated, the problem has been solved, usually taking some approximations, such as assuming that the potential well is perfectly triangular in shape. However, the wells cannot be assumed to be of infinite height, since in our case $\Delta E_c \approx 0.3\,\text{eV}$. Detailed calculations also allow calculation of the average width of the well (40–80 Å), the electron concentration per unit area, $n_s \approx 10^{12}\text{cm}^{-2}$, and the energy ε_1 of the first level $\approx 0.04\,\text{eV}$.

Evidently, if we want to construct FET based on AlGaAs–GaAs heterojunctions, source and drain contacts have to be deposited as shown in Figure 5.1 for the MOSFET. By applying differences of potential to the gate, the number of carriers in the channel, and therefore its conductance, can be controlled as in the case of the MOSFET (Section 5.2). These transistors are called modulation-doped field effect transitor (MODFET), from modulation doping, or high electron mobility transistor (HEMT). In Section 9.2 we will consider in detail MODFETs, which are used in many high-frequency applications due to the very high electron mobility of the electrons in the channel.

5.3.2.   SiGe strained heterostructures

SiGe heterojunctions did not attract too much attention at first because of the large lattice constant difference between Si and Ge (Figure 4.9), which amounts to about 4%. This means that the layers grow strained over the substrate and that a critical thickness should not be surpassed (Section 4.7) otherwise the structure breaks off. It is also true that, since the energy gaps of silicon ($E_g = 1.12\,\text{eV}$) and germanium ($E_g = 0.66\,\text{eV}$) are fairly small, the height of the barriers which appear at the interface should always be small. In spite of these difficulties, SiGe heterostructures have found interesting applications in several fields such as high frequency transistors (Chapter 9) and IR photodetectors (Chapter 10). Since SiGe are strained, the degeneracy of the heavy and light hole bands is lifted (Section 4.8), and the band structures show similar features to those shown in Figure 4.12.

Figure 5.5 shows two typical examples of SiGe heterostructures. In Figure 5.5(a) the substrate is <001> Si ($E_g = 1.17\,\text{eV}$) and the strained active layer $Si_{0.7}Ge_{0.3}$ ($E_g = 0.78\,\text{eV}$). In this case the conduction band offset is rather small, in contrast to the valence band offset. This situation allows the formation of a 2D hole gas in the SiGe alloy, with electron mobilities around $2\,\text{m}^2\text{V}^{-1}\text{s}^{-1}$, i.e. about half the value found for electrons in a typical MOSFET (Section 5.2). In Figure 5.5(b) the situation is reversed and the strained layer is Si. In this case the discontinuity in the conduction band is fairly large and the electrons form a 2D gas, with free motion in the plane of the interface. The silicon effective mass corresponding to this motion is the low transversal one ($m_T^* \approx 0.19 m_0$), therefore yielding a high mobility of around $20\,\text{m}^2\text{V}^{-1}\text{s}^{-1}$, several times higher than the one corresponding to the MOSFET.

SiGe heterostructures have also found an important application in the field of bipolar silicon transistors which will be considered in more detail in Section 9.3. One way to improve the efficiency of a bipolar transistor is to use a narrow-bandgap material for

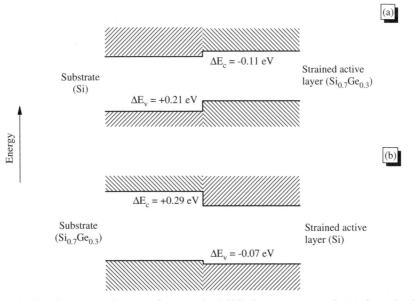

Figure 5.5. Simplified band diagram of two strained SiGe heterostructures. In (a) the active layer is SiGe, while in (b) the SiGe acts as a substrate.

the base region, which improves the efficiency of the Si emitter region. In this case the advantage stems from the reduced values of the SiGe bandgap alloys, in comparison to silicon. It is important to mention that the strain that appears in the heterojunction also contributes to the decrease of the bandgap. In addition, the large bandgap offset allows the fabrication of a highly doped, low resistivity, base material, which extends the performance of silicon transistors to much higher frequencies.

5.4. QUANTUM WELLS

5.4.1. Modulation-doped quantum well

It is often desirable, especially in the case of multiple quantum wells, that the individual wells are approximately symmetric and square-shaped, instead of being triangular like the simple modulation-doped heterojunction studied in the previous section. Let us now consider that we build a symmetric well by facing two AlGaAs–GaAs heterojunctions opposing each other like in Figure 5.6(a). The wide gap semiconductor material $Al_x Ga_{1-x} As$ is located at the ends and the GaAs in the middle. Imagine next that the distance between the two interfaces is made sufficiently small. Then the resulting well

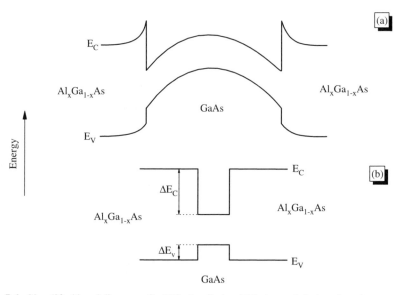

Figure 5.6. Simplified band diagram of a AlGaAs–GaAs–AlGaAs modulation-doped quantum well.
(a) Case of a wide well; (b) case of a narrow well.

(Figure 5.6(b)) for electrons and holes would be almost square with a barrier on each side of the same height as in Figure 5.6(a).

It is important to remark, as in Section 5.3.1, that the volume of material inside the well is free of ionized donors (located in the AlGaAs material). Therefore the electrons inside the well, which originated at the neighbouring AlGaAs donor-material, can move into the GaAs region or channel with very high mobility. As in the case of the modulation-doped heterojunctions, MODFET high-frequency transistors can be fabricated if appropriate source and drain contacts are deposited.

Quantum well structures with either high or low mobility for electrons can be fabricated by introducing a controlled amount of impurities. A double quantum well structure with high and low mobilities constitutes the base of the velocity-modulation transistors. In these transistors, the switching from one state to the other is controlled by an electric field transverse to the layers which redistributes the amount of charged electrons and therefore the current in either wells. Velocity-modulation transistors can be operated at very high frequencies.

It is quite difficult to find accurate equations for the wave functions and energy levels of electrons and holes in the wells, since the potential at the bottom, instead of being flat, is influenced by the variations at the interfaces and the problem has to be solved by numerical methods. However, in order to describe the general behaviour, we can use perturbation theory and consider the potential variations as a perturbation. The perturbation potential should be symmetric, and thus an even function of z, which leads to the important conclusion that only states with the same parity can be mixed, as we will see in more detail, when we study optical transitions in quantum wells (Chapter 8). For instance, if we assume that only three levels exist in the well, the first state can only be coupled with the third. Evidently, the occupation of levels depends on the electron concentration in the well, and for low concentrations usually the first level is the only one occupied.

5.4.2. Multiple quantum wells (MQW)

The signal provided by a single quantum well is usually too small to be used in the solid state devices. Therefore, it is often necessary to use an array of quantum wells, especially in optoelectronic devices, such as photodetectors. These structures are called multiple quantum wells (MQW) and are formed by several single quantum wells. If the wells for electrons and holes are located in the same space location, the MQW is called *Type I* (Figure 5.7(a)), while the name *Type II* is used when the corresponding wells are located alternatively as in Figure 5.7(b).

In a MQW system it is assumed that there is no interaction between neighbouring quantum wells, because the barriers separating the wells are thick enough, usually more than about 10 nm. However, if the energy barriers between consecutive wells are thin enough, the wells will be coupled to each other by tunnelling effects. As we will see in the next section, the discrete energy levels of the quantum wells are then transformed into energy bands. In this case, the system of MQWs is called a superlattice and the energy spectrum shows very interesting new features.

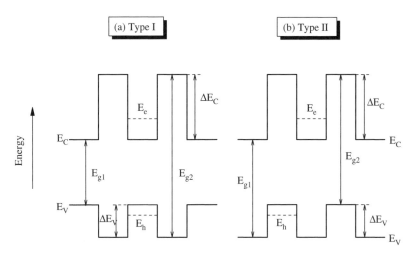

Figure 5.7. (a) Type I MQW; (b) Type II MQW.

In optoelectronics, MQWs are frequently used and are made of about 50 single wells. In this case, if we consider that the periodicity (thickness of the well plus barrier) is about 20 nm, then the total thickness of the array approaches 1 μm. This value is usually taken as the limit in thickness; otherwise radiation would be strongly absorbed before reaching the inner quantum wells. It should also be observed that the spectrum of the total optical signal does not necessarily coincide with that of a single quantum well since the thickness of each well is not exactly the same for all (consider for instance the percentage variation in the width of a quantum well when the thickness varies by just one monolayer). Consequently, the breadth of the MQW energy levels can be used as an index of its uniformity in thickness.

Figure 5.8 shows the band structure of a typical MQW for applications in IR photodetectors or in electro-optic switching modulators under an applied electric field. Photodetectors for the IR usually work at wavelengths in the 10 μm range. If we were to detect this long wavelength radiation using bulk semiconductors, a gap of only about 0.1 eV would be necessary. Therefore it constitutes a better option to use intersubband detection in quantum wells. In the design of these detectors we have the freedom to adjust the well height ΔE_c (by changing the value of x in $Al_xGa_{1-x}As/GaAs$), as well as the values of energy levels through the thickness of the well material. As shown in Figure 5.8, the levels are calculated in such a way that the second level is only a little below the bottom edge of the conduction band of the wide gap material ($Al_xGa_{1-x}As$). In addition, the semiconductor doping is chosen so that the $n = 1$ level is full and the $n = 2$ empty. Taking these considerations into account, when a photon of energy $\hbar\omega$, equal to $E_2 - E_1$ impinges, electrons will be liberated and drift along the z-axis by the action of a weak applied electric field.

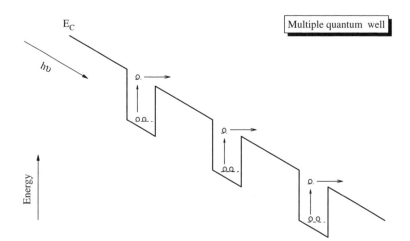

Figure 5.8. Band structure of a typical MQW for IR photodetectors, under an applied electric field.

5.5. SUPERLATTICES

5.5.1. The concept of a superlattice

The concept of the superlattice (SL) was proposed by Leo Esaki and coworkers in the late 60s and later fabricated by molecular beam epitaxy (MBE) techniques. Superlattices can be considered one of the most important man-made or artificial materials. A super-lattice consists of a periodic set of MQW in which the thickness of the energy barriers separating the individual wells is made sufficiently small. As the barriers become thin-ner, the electron wave functions corresponding to the wells overlap due to the tunnelling effect. As a consequence, the discrete energy levels of the wells broaden and produce energy bands, in a similar way as happens with the states of the individual atoms when they are arranged in a crystal lattice. The most singular aspect of a superlattice consists of introducing at will a new periodicity d in the material, which is equal to the breadth of the well a, plus the thickness of the barrier b. Typical thicknesses for a and b could be 4 and 2 nm, respectively. An accurate control over these small thicknesses can only be achieved by techniques for thin film deposition such as molecular beam epitaxy or metal organic chemical vapour deposition.

In order to study the origin of the band structure of superlattices, let us consider first the overlapping between the electron states for a simple two-well system. This is already a familiar problem, because from a quantum mechanical point of view it is formally similar to the case of the diatomic molecule. Figure 5.9(a) shows two neighbouring iden-tical quantum wells and corresponding wave functions of what is known as the double coupled quantum well system. The solution for this problem is based on perturbation theory in quantum mechanics. According to it, each original level, say E_1, of the isolated wells splits into two, with energies

$$E = E_1 \pm |V_{12}| \tag{5.4}$$

shown in Figure 5.9(b). V_{12} is given by the overlap integral

$$V_{12} = \int_{-\infty}^{+\infty} \psi_1^* V(z) \psi_2 dz \tag{5.5}$$

The two resulting levels are separated in energy by $2|V_{12}|$, where the magnitude of V_{12} of Eq. (5.5) is an indication of how much one well can influence the energy states of the neighbouring one, hence the name overlap integral.

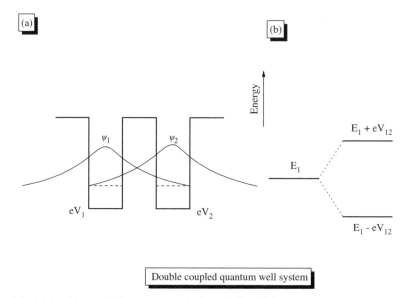

Figure 5.9. (a) Double coupled quantum well; (b) splitting of the energy levels.

5.5.2. *Kronig–Penney model of a superlattice. Zone folding*

In order to determine the electronic band structure of superlattices, one can proceed as in solid state physics if we assimilate the superlattice to the crystal lattice and the quantum well potentials to the atom potentials in the crystal. Consequently, we can study the superlattice band structure from the point of view of two approximations: the Kronig–Penney model and the tight binding approximation, as we did for crystalline solids in Chapter 2.

It is interesting to remark that as early as in 1931, Kronig and Penney established a model [3] for a solid in which the periodic potential seen by the electrons was precisely that of the square type shown in Figure 5.10 for a superlattice potential. This periodic

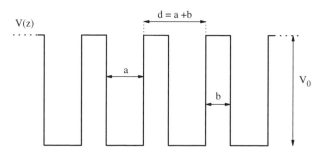

Figure 5.10. Scheme of the periodic potential of a superlattice.

one-dimensional potential is characterized by the following parameters: well thickness a, barrier thickness b, and barrier height V_0. The spatial periodicity is $d = a + b$. Even if the periodic potential of Figure 5.10 is much simpler than the real one, the Kronig–Penney model yields very interesting results related to the structure of the bands, the forbidden zones, size of the gaps, etc. In the well region ($0 < z < a$), $V = 0$, and the wave function is, according to Eq. (2.11).

$$\psi(z) = Ae^{ik_0 z} + Be^{-ik_0 z} \tag{5.6}$$

with

$$k_0^2 = \frac{2mE}{\hbar^2} \tag{5.7}$$

Due to tunnelling, the wave function extends inside the energy barrier of height V_0 and thickness b. Therefore, if $-b < z < 0$,

$$\psi(z) = Ce^{qz} + De^{-qz} \tag{5.8}$$

where the wave vector and the energy are related by

$$V_0 - E = \frac{\hbar^2 q^2}{2m} \tag{5.9}$$

From the conditions that both the wave functions and their derivatives are continuous at $z = 0$ and $z = a$ (the origin in z is taken at the left barrier of the well), we get after operating

$$A + B = C + D \tag{5.10}$$

$$ik_0(A - B) = q(C - D) \tag{5.11}$$

According to the Bloch theorem, expressed in the form of Eq. (2.40), we can relate the wave functions at two different locations by

$$\psi(a) = \psi(-b)e^{ik(a+b)}$$

where k is the wave vector corresponding to the Bloch wave functions. Applying this equation to the wave functions corresponding to the well and the barrier regions, Eqs (5.10) and (5.11), respectively, we have:

$$Ae^{ik_0a} + Be^{-ik_0a} = (Ce^{-qb} + De^{qb})e^{ik_0(a+b)} \qquad (5.12)$$

$$ik_0(Ae^{ik_0a} - Be^{-ik_0a}) = q(Ce^{-qb} - De^{-qb})e^{ik_0(a+b)} \qquad (5.13)$$

The four Eqs (5.10)–(5.13) for the amplitudes A, B, C, and D have a solution only if the determinant of the coefficients equals zero. After some calculations, one gets the important relation [4]:

$$\frac{q - k_0^2}{2qk_0} \sin k_0a \sinh qb + \cos qa \cosh qb = \cos q(a + b) \qquad (5.14)$$

It is not difficult to solve this equation numerically. Let us assume the simple case of $a = b$ and further that the effective mass of the electron is the same in the well and barrier materials. For the case of GaAs–AlGaAs and $E < V_0$, solution of Eq. (5.14) gives the values of allowed and forbidden energies, which are represented in Figure 5.11 for a superlattice with a band discontinuity of 0.3 eV and an effective mass, $m_e^* = 0.067m_0$ [5]. The figure shows that for values of a larger than about 10 nm, the electron energies are well defined and correspond to the individual quantum wells. However, when the barrier width is smaller than about 6 nm, bands as well as forbidden zones arise.

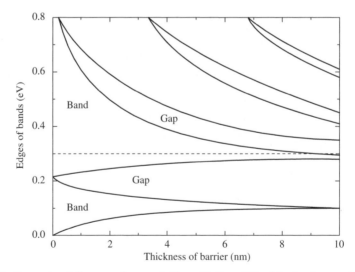

Figure 5.11. Energy band diagram of a superlattice with the width of the barrier equal to the width of the wells. After [5].

It is evident from Figure 5.12 that the general features of the *E–k* relationship in a superlattice, within the Kronig–Penney model, are very similar to the case of electrons in solids (Section 2.5) if we replace the bulk lattice constant by the larger period of a superlattice. The free electron parabola, therefore, breaks down into several bands and gaps at the edges of the Brillouin zones $k = \pm n\pi/d$, as shown in the extended *E–k* diagram of Figure 5.12. Next, the portions of the bands can be translated to the reduced zone $(-\pi/d \leq z \leq +\pi/d)$. Notice that everything occurs as if the superlattice potential folds the quasi-free energy band of a solid into the centre of the reduced zone, as it can be appreciated if we make the representation in the first Brillouin zone. Since, usually $d \gg a$, the breadth of the bands and gaps in a superlattice are much smaller, often receiving the names of minibands and minigaps. This band folding procedure is typical of superlattices and is called *zone folding* since it implies that the pieces of the band in the extended representation are zone-folded into the smaller zone with values of *k* smaller than $2\pi/d$.

The zone-folding effect has important consequences in the direct or indirect character of semiconductor structures. Figure 5.13 represents the band diagram of a typical indirect gap semiconductor with the minimum of the conduction band at the zone edge. Suppose next that we construct a superlattice by alternating monolayers of two semiconductors with similar electronic properties and well-matched lattice constants, but with one of

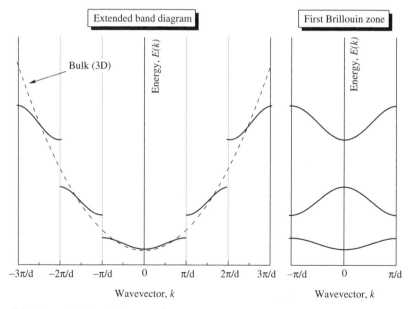

Figure 5.12. Extended band diagram of a superlattice in reciprocal space (left). On the right, the band diagram is represented in the reduced first Brillouin zone.

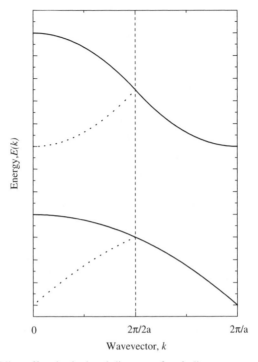

Figure 5.13. Zone-folding effect in the band diagram of an indirect gap semiconductor.

them with the conduction band as in Figure 5.13. As indicated by the dotted line in the figure, the minimum of the conduction band is translated to $k = 0$, according to the folding procedure. The resulting semiconductor structure should, therefore, show a direct gap at $k = 0$ of the same energy value as the indirect one.

Based on the zone folding concept, one can fabricate superlattices of GaAs and AlAs with direct and indirect gaps, respectively, which show quasi-direct optical transitions. Another example is that of the $Si_n Ge_m$ superlattices based on SiGe heterostructures (Section 5.3.2). In these superlattices, which are constructed usually symmetrical ($n = m$), the minimum of the Si conduction band close to the band edge can be brought close to $k = 0$ by successive zone-folding. Evidently, the higher the number of monolayers (n), the behaviour of the superlattice would better resemble that of a direct bandgap semiconductor structure.

It is interesting to note that due to the small widths of the minigaps and minibands in superlattices, as well as the quasi-direct type of transitions which might show, they find many applications in infrared optics. In addition they can show new interesting properties like Wannier–Stark localization and Bloch oscillations (Section 8.5).

5.5.3.   Tight binding approximation of a superlattice

In this section we will deduce a series of properties of the band diagram of a superlattice from the tight binding approximation for solids (section 2.5.2), in a similar manner to the previous section by using the Kronig–Penney model. For this purpose we consider the superlattice as a set of N quantum wells along the z-direction that are weakly coupled, in analogy with the potentials felt by electrons in solids. For instance, the Bloch wave function in the ground state of the superlattice $\psi_{g.s.}$ should be a linear combination of the wave functions of each quantum well $\psi(z - nd)$ of potential energy $V(z - nd)$, where we are supposing that each well is centred at locations $z = nd$

$$\psi_{g.s.} = \frac{1}{\sqrt{N}} \sum_n e^{iqnd} \psi(z - nd) \tag{5.15}$$

Proceeding as in the tight binding approximation, the Hamiltonian for the Schröedinger equation for the wave function can be written as $H_0 + H_1$, where H_0 is the Hamiltonian for one isolated well and H_1 is the perturbation, in this case, the potential due to all the other wells. Solving the perturbation problem, within the nearest neighbour approximation, one gets following the same method as for the derivation of Eq. (2.49):

$$E(q) = E_0 + s + 2t \cos qd \tag{5.16}$$

According to Eq. (5.16) the shape of the band follows a sinusoidal function (Figure 5.14(a)), which is similar to Figure 2.6 for bands in solids [6]. Note also from Eq. (5.16) that the band width is $\Delta E = 4|t|$, i.e. depends on the transfer integral t, which takes into account the coupling between nearest neighbours and depends on the super-lattice parameters. Figure 5.14(b) shows the dependence of the bandwidth on the barrier thickness. As expected, when the barriers between wells get thicker, and therefore $t \approx 0$, we should have the same result as for multiquantum wells with single levels for the energy.

One interesting point that should be emphasized is related to the periodicity of the superlattice in the z-direction, since in reality superlattices are three-dimensional struc-tures. Therefore, the expression of the total energy should also take into account the kinetic energy of electrons for their motion along the (x, y) planes. The total energy of the electrons in the i subband should be equal to the kinetic term plus the energy corresponding to the motion along the superlattice direction and given by equation

$$E(k, q) = \frac{\hbar^2 k_{\|}^2}{2m_{\|}} + E_0 + s + 2t \cos qd \tag{5.17}$$

where the subindexes of k and m refer to the bi-dimensional motion through the interface. The DOS function can be calculated as we did for a quantum well in Section 4.2, where we

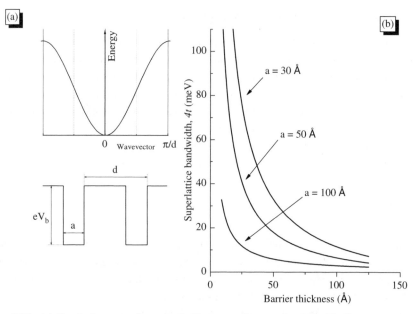

Figure 5.14. (a) Band diagram of a superlattice according to the tight binding approximation;
(b) dependence of the width of the energy bands with the barrier thickness. After [6].

obtained the factor $m_{\parallel}/\pi\hbar^2$ for each band. Performing in the above expression the integral over q from $-\pi/d$ to $+\pi/d$ we obtain, after some straightforward calculations [4], the following expression for the DOS function:

$$n_{SL} = N\frac{m_{\parallel}}{\pi^2\hbar^2}\cos^{-1}\left(\frac{E_i - E_{oi} - s_i}{2t_i}\right) \qquad (5.18)$$

Figure 5.15 shows the DOS function for a superlattice, together with the one for a quantum well [5]. As a reference, the 3D-DOS function is also represented. Note that at $E = E_n$, i.e. the values corresponding to the subband energies, the superlattice DOS function has a value which equals half the value for a single quantum well. It should also be remarked that as the value of the transfer integral t increases, the deviations to the quantum well spectrum also gets larger.

5.5.4. *nipi superlattices*

The superlattices treated so far consisted of a periodic array of individual quantum wells (Figure 5.10). They were obtained by alternating two materials with different gaps. As proposed by Esaki and Tsu, an alternate way to produce a superlattice would consist

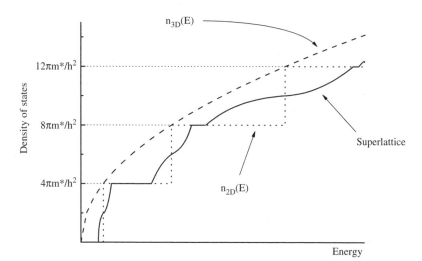

Figure 5.15. Density of the states functions for a superlattice.

of the periodic arrangement of p–n homojunctions. In this case, there would be no band discontinuities since the p or n character of a material, for instance, silicon, does not change the magnitude of the energy gap. This type of structure is called a *doping or nipi superlattice*. In this notation, the letter *i* stands for intrinsic since between p and n regions there is always an intrinsic zone, however small, separating them. On other occasions an extra intrinsic region is added purposely. In contrast to the quantum well superlattices, where the electric potential space modulation along the growth direction was caused by the conduction and valence band discontinuities, now the modulation is produced by ionized donors and acceptors, which create the p–n space charged region around the interfaces.

Figure 5.16(a) represents an imaginary situation for the n and p regions, where the respective impurity donors and acceptors have not been ionized yet. However, when the unions are established, the donors and acceptors get positively and negatively charged, respectively. As a consequence the potential energy differences originated by the space charge causes the bending of E_c and E_v as represented in Figure 5.16(b). Taking into account the electrostatics of the p–n junctions, it is evident that the amplitude of the modulation E_M depends on the dopant concentration (N_D, N_A). The magnitude of the superlattice gap or energy difference between E_c and E_v is evidently, from Figure 5.16(b) equal to the gap of the basic material forming the homojunction E_G minus E_M, i.e.

$$E_{SL} = E_G - E_M \tag{5.19}$$

Obviously, the gap can be tailored by adjusting the dopant concentrations. From Figure 5.16(b) it can also be observed that the potential wells for electrons are located at

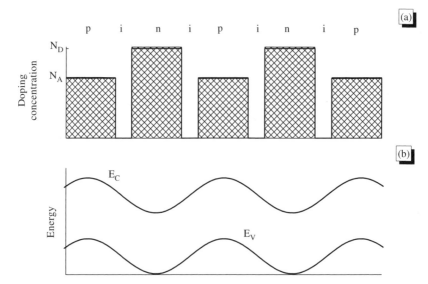

Figure 5.16. (a) Doping concentration of a nipi superlattice; (b) bottom edge of conduction band
and upper edge of the valence band for a nipi superlattice.

the n regions. Around the minimum of the well, the potential can be approximated by a
parabolic one (Section 4.4.1). Within the harmonic oscillator approximation, the energy
of the electrons in the conduction band is quantized, according to Eq. (4.11). From the
solution of Poisson's equation for the p–n junction, it is easy to calculate (Problem 5.5)
the values E_n of the quantized energy levels:

$$E_n = \hbar e \left(\frac{N_D}{\varepsilon m_e^*}\right)^{1/2} \left(n + \frac{1}{2}\right), \quad n = 0, 1, 2, \dots \tag{5.20}$$

which depends on material parameters such as the effective mass, dielectric constant, and
impurity concentration. Some additional properties of nipi superlattices are: (i) The mobilities are lower than in compositional superlattices because the carriers are not separated
in space from the impurities. (ii) Optical absorption is in general weaker than in compositional superlattices, due to the fact that transitions from valence band levels to conduction
band ones have lower probability as a consequence of the indirect space character of the
superlattice, since electrons and holes are located at different places in real space. (iii) The
carriers show a long lifetime as a consequence of their separation after they are generated.
(iv) The nipi superlattices show interesting applications in optical computing based on the
quantum confined Stark effect (QCSE) (Section 8.4).

REFERENCES

[1] von Klitzing, K., Dorda, G. & Pepper, M. (1980) *Phys. Rev. Lett.,* **45,** 494.
[2] Hamaguchi, Ch. (2001) *Basic Semiconductor Physics* (Springer, Berlin).
[3] Kronig, R.de L. & Penney, W.G. (1931) *Proc. Roy. Soc. A,* **130**, 499.
[4] Basu, P.K. (1997) *Theory of Optical Processes in Semiconductors* (Clarendon Press, Oxford).
[5] Esaki, L. (1983). In *Recent Topics in Semiconductor Physics,* Eds. Kamimura, H. and Toyozawa, Y., 1–71 (World Scientific, Singapore).
[6] Bastard, G. (1993) *Acta Electron.,* **25**, 147.

FURTHER READING

Davies, J.H. (1998) *The Physics of Low-Dimensional Semiconductors* (Cambridge University Press, Cambridge).
Ferry, D.K. & Goodnick, S.M. (1997) *Transport in Nanostructures* (Cambridge University Press, Cambridge).
Hamaguchi, Ch. (2001) *Basic Semiconductor Physics* (Springer, Berlin).
Jaros, M. (1989) *Physics and Applications of Semiconductor Microstructures* (Clarendon Press, Oxford).
Kelly, M.J. (1995) *Low-Dimensional Semiconductors* (Clarendon Press, Oxford).
Mitin, V.V., Kochelap, V.A. & Stroscio, M.A. (1999) *Quantum Heterostructures* (Cambridge University Press, Cambridge).
Murayama, Y. (2001) *Mesoscopic Systems* (Wiley-VCH, Weinheim).
Schäfer, W. & Wegener, M. (2002) *Semiconductor Optics and Transport Phenomena* (Springer, Berlin).
Weisbuch, C. & Vinter, B. (1991) *Quantum Semiconductor Structures* (Academic Press, Boston).

PROBLEMS

1. **MOSFET.** (a) Make an estimation of the value of the electric field in an n-channel MOSFET-Si transistor assuming a gate voltage V_G of 4 V and an oxide thickness of 50 nm (Assume $\varepsilon_{ox} = 3.6$, $\varepsilon_{Si} = 12$). *Hint*: from the value of the electric field in the oxide, calculate the electric field at the interface applying the boundary condition on the displacement vector. (b) Calculate the energy for the ground level. *Hint*: assume the shape of the well is triangular.

2. **Modulation-doped heterostructure**. Consider a modulation-doped AlGaAs/GaAs heterostructure with a carrier density n_{2D} equal to 10^{12} cm^{-2}. What is the position of the Fermi level at room temperature, assuming that the electrons are all located in the first subband? *Hint*: show first that if E_1 is the energy of the first subband, then

$$n_{2D} = \frac{m_e^* kT}{\pi \hbar^2} \ln \left[1 + \exp \left(\frac{E_F - E_1}{kT} \right) \right]$$

3. **Minibands in superlattices**. Consider an AlGa$_{1-x}$As/GaAs superlattice with $x \approx 0.3$, a well width $a = 10$ nm and barrier width of $b = 2.2$ nm and a barrier height of $V_0 = 0.25$ eV. From the theory of the Kronig–Penney model developed in Section 5.5.2: (a) Find the widths of the minibands and minigaps that can exist within the barrier. (b) Maintaining fixed value of the width ($a = 10$ nm), plot the energy of the first ($n = 1$) conduction miniband and heavy-hole miniband as a function of the barrier width b between 0 and 5 nm.

4. **Si/SiGe superlattices**. Calculate the period of a Si/SiGe superlattice, so that the band minimum of Si, which occurs at wave number of approximately $k \approx 0.8\pi/a_0$, (a_0 is the lattice constant) be brought to $k = 0$. *Hint*: according to the zone folding concept of Section 5.5.2, observe that the superlattice can behave optically as a direct gap bulk semiconductor.

5. **nipi superlattices**. Consider a gallium arsenide nipi superlattice with equal $N = N_D = N_A$ donor and acceptor concentrations of value 5×10^{17} cm^{-3}. (a) Calculate the separation between levels in the conduction and valence bands. (b) Show that the value of the effective bandgap is given by

$$E_g = E_{bulk} + E_{e1} + E_{h1} - 2V_0$$

where V_0 is the amplitude of the periodic potential. Show that, if $N_A = N_D = N$, the amplitude of the periodic potential in the superlattice is given by

$$V_0 = \frac{e^2}{2\varepsilon} N z_0^2$$

where z_0 is the depletion layer width. (d) Calculate V_0 and the effective bandgap assuming the previous concentration doping and taking $z_0 = 20$ nm.

Chapter 6

Electric Field Transport in Nanostructures

Chapter 6

Electric Field Transport in Nanostructures

6.1. INTRODUCTION

In previous chapters, we have studied the formation of quantum wells at interfaces between semiconductors of different gaps. Conduction band electrons in these quasi-2D wells behave almost as free carriers for their motion along planes parallel to the interfaces of the well. For this kind of transport, usually called *parallel transport*, a semiclassical approach, somewhat similar to the 3D bulk case, is usually valid. The main differences which arise are related to the characteristics of the density of states function, and to those electron scattering mechanisms which are peculiar to low-dimensional systems. However, in the case of transport through the potential barriers at the interfaces, which is known as *perpendicular transport*, the mechanisms involved are completely different to those present in the bulk and are mainly based on the quantum tunnelling effect.

In this chapter, we study the transport properties due exclusively to the action of electric fields and we will postpone to the next chapter, the study of these properties due to the joint action of electric and magnetic fields. We will see that entirely new quantum effects arise when transport in a low-dimensional semiconductor is studied under the action of magnetic fields, such as for instance, the quantum Hall effect and the Aharonov–Bohm effect. However, we would like to remark that, even under the sole action of electric fields, transport in nanostructures also shows some new unexpected effects such as quantized conductance, Coulomb blockade, etc.

6.2. PARALLEL TRANSPORT

Electronic transport in 2D quantum heterostructures, parallel to the potential barriers at the interfaces, can be treated following a semiclassical approach as in the bulk case, if we take into consideration that there are additional electron scattering mechanisms (e.g. scattering due to interface roughness) and that we are dealing with a low-dimensional system. Parallel transport in nanostructures was investigated first in the conduction of electrons along the channel of MOSFET structures. Later, it received a great boost with the fabrication in the 1970s of MODFETs based on modulation-doped quantum heterostructures. Common to both devices is the fact that electron motion takes place in a region free of charged dopants, and therefore, electrons can reach very high mobilities.

6.2.1. Electron scattering mechanisms

The main scattering mechanisms for parallel transport in semiconductor nanostructures are due, as in the case of bulk samples, to phonons and impurities (neutral and ionized). In addition, there are other mechanisms, characteristic of nanostructures, such as scattering due to interface roughness. Below we treat each of these scattering mechanisms separately.

(i) Electron–phonon scattering

Calculations performed on electron–phonon scattering mechanisms in low-dimensional semiconductors show that the results are somewhat similar to the bulk case. In this sense, the phonon scattering mechanism is the predominant one for temperatures higher than about 50 K. However, when the width a of the quantum wells becomes very small, the role of acoustic phonons is different, and usually more important, than in the 3D case, as a consequence of the non-translational invariance in the perpendicular direction. This can be appreciated, for instance, in a 2D quantum well for which the uncertainty in the perpendicular component of the momentum should be $\geq h/a$. Hence, contrary to the bulk case, where acoustic phonons have well defined momenta, in the case of electron–phonon scattering in very narrow quantum wells, the phonon momentum is not conserved. As the uncertainty in momentum increases, the number of electron–phonon scattering mechanisms also increases and for this reason phonon scattering becomes very considerable in low-dimensional semiconductors.

The case of optical phonons is quite different from the bulk one, especially for nanostructures of strongly polar materials such as III-V compounds. The interaction is especially strong in quantum wells (Section 5.4) when there is no overlapping between the optical phonon energy bands of the well semiconductor (e.g. GaAs) and the barrier semiconductor (e.g. AlGaAs). If this is the case, the contributions to phonon scattering of the confined optical modes and those associated to the interfaces become much more important than the contribution of bulk-like optical phonons.

(ii) Impurity scattering

As for bulk samples, ionized and neutral impurity scattering constitutes the largest contribution to scattering in low-dimensional semiconductors at low temperatures. The main difference between scattering events in a bulk or in a 2D system is that, for parallel transport, the location of the impurities is often separated from the 2D plane in which electrons move. In modulation-doped heterostructures (Figures 5.4 and 5.6), the charged donors are located in the AlGaAs, while electron motion takes place in a separated region in the GaAs parallel to the interface. Similarly, in a MOS structure (Figure 5.1), electrons move within the inversion channel, which is separated from impurities located in the thin gate oxide.

For the calculation of impurity scattering in MODFET quantum heterostructures (Section 5.3.1) some simplifying assumptions are usually made, such as δ-doping, i.e. the ionized impurities are supposed to be located in a 2D plane at a distance d of the electron channel, and that the electrons in the channel which participate in the scattering events are those with energies very close to the Fermi level. Furthermore, it is also assumed that the concentration of impurities is not too high, so that each charged impurity interacts independently with the carriers. With these assumptions, is not difficult to reach the conclusion (see for instance, Ref. [1]) that the mobility of the carriers increases as d^3. However, there is an optimum value of d, because if d is too large, the concentration of electrons in the channel diminishes significantly, as a consequence of the decrease in the electric field, and the transconductance of MODFETs is greatly reduced.

(iii) Surface roughness scattering

Interface scattering is due to the interaction of electrons with a roughened surface, in contrast to an ideal perfect flat surface for which this interaction would be elastic. Real interfaces have a roughness at the atomic level, which produces non-specular reflections of carriers, and therefore, a loss of momentum contributing to relaxation mechanisms. Interface scattering has been studied for a long time, due to its important role in transport along thin films, but modern quantum theories in low-dimensional systems were not developed until the 1980s. The role of interface scattering for parallel transport in modulation-doped heterostructures is not very important, due to the high perfection of the interfaces when growth techniques such as molecular beam epitaxy are used. In this case the surfaces are practically flat, with few steps of the size of a monolayer.

In the case of MOS structures, interface scattering becomes more important since the oxide is grown thermally and the interface is not as perfect as in the modulation-doped heterostructure. The contribution of interface scattering in MOS structures depends on the quantum well width. In effect, as the width decreases the electron wave function penetrates deeper into the oxide-semiconductor potential barriers, i.e. the electrons are more exposed to the interface roughness and the corresponding scattering increases. This is the reason behind the observed decrease of mobility with applied gate voltage. Anyway, roughness scattering, like impurity scattering, only becomes significant at temperatures low enough for phonon scattering to be negligible. Finally, in the case of narrow quantum wires, the contribution of interface scattering is almost one order of magnitude larger than in 2D systems. This is especially so when the boundaries of the wires are defined by lithographic techniques, in which case surface roughness can be a limiting factor for the mobility, even at room temperature.

(iv) Intersubband scattering

Let us consider a 2D electron system confined in a potential well as in the case of a modulation-doped quantum heterojunction or the MODFET (Section 5.3.1). It is evident that for large electron concentrations in the well, the levels with energies higher than the first one E_1 will start to become filled. Imagine a situation in which the electron concentration is high enough so that the Fermi level E_F just crosses the quantized level corresponding to $n = 2$. Then, electrons with energies around E_F can either undergo an intraband scattering transition within the subband $n = 2$ or an interband transition between subbands $n = 1$ and $n = 2$. Therefore, these electrons have two possible scattering channels and the total scattering probability should increase. As a consequence, the electron mobility should become smaller. This effect can be generalized to other subbands. In summary, as the electron concentration in a quantum well increases, additional scattering channels start to contribute to the overall scattering rate, and the mobility of the 2D electron gas decreases. This effect is even more acute in 1D systems for which the density of states diverges at those values of energy coinciding with the quantum levels (Section 4.5).

The effect of subband scattering in the mobility has been studied by Störmer et al. in 1982 in AlGaAs/GaAs modulation-doped heterojunctions [2] to which a third terminal (the gate) was added, like in the MODFET, in order to control the electron concentration in the well. As Figure 6.1 clearly shows, the onset of a second channel allows intersubband scattering between the $n = 1$ and $n = 2$ subband. At a given gate voltage, the Fermi level reaches the $n = 2$ level and the corresponding subband can host scattered electrons, thus decreasing mobility at a given voltage interval.

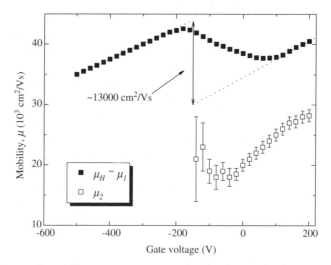

Figure 6.1. Decrease in mobility as a consequence of subband scattering between subbands $n = 1$ and $n = 2$. After [2].

6.2.2. Experimental data on parallel transport

Figure 6.2 shows the large improvement, experienced during a 12-year period by the electron mobility for parallel transport at low temperatures in GaAs-based nanostructures [1], like for instance MODFETs. There are several reasons for this success. The main reason, as pointed out before, is due to the physical separation between dopants and carriers in modulation-doped heterostructures. In order to make this separation even more effective, a semi-insulating layer, called a spacer, is added between the donor layer and the 2D electrons in the conducting channel. This spacer is especially effective at low temperatures for which the impurity–electron scattering mechanism becomes predominant. Another reason for the large increase in the electron mobility is the high purity of the bulk material, caused by the improvement in the growth techniques of III-V materials, as can be appreciated in Figure 6.2 for the "clean bulk" curve, which for $T \geq 100$ K shows excellent results. The improvements in the purity of the layer constitute another important factor and are due to the ultrahigh vacuum and gas purity characteristics of growth techniques such as molecular beam epitaxy. As the temperature approaches 100 K and gets close to room temperature, the dominant scattering mechanisms are due to phonons, especially optical phonons in the case of highly polar substances like GaAs (Section 6.2.1 (i)).

As we should expect, the mobility of electrons in a silicon MOSFET should be much lower than in a MODFET. The mobility in modulation doped AlGaAs/GaAs can reach values as high as 10^7 cm^2 V^{-1}s^{-1}, while in a MOSFET the values are about three orders of magnitude lower (Figure 6.3) [3]. There are several reasons for this. First, the effective

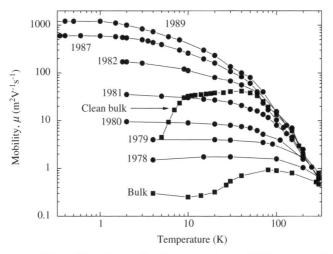

Figure 6.2. Increase of the mobility for parallel electron motion in III-V compound heterojunctions (the value for bulk GaAs is also shown). After [1].

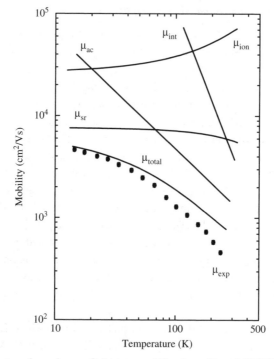

Figure 6.3. Temperature dependence of electron mobility in a silicon MOSFET. After [3].

mass of electrons in Si is much higher than in GaAs. Second, the effect of impurity scattering in a Si MOSFET, caused by charges and impurities in the oxide and the interface, is larger than in the case of AlGaAs/GaAs, where the separation of charged impurities and carriers is very effective. Third, as can be appreciated in Figure 6.3 the effect of surface roughness at low temperatures becomes the dominant scattering factor. This was expected because, as we remarked in Section 6.2.1(iii), the silicon–oxide interface, grown thermally, is not as perfect as the AlGaAs/GaAs interface produced by sophisticated techniques such as MBE.

Under the action of very high electric fields ($\geq 1\,\mathrm{MV\,cm^{-1}}$), surface roughness scattering dominates over all other mechanisms in limiting the value of the electron mobility. In this respect, it is also interesting to mention that in quantum wires, surface roughness scattering mechanisms, produced as a consequence of present patterning techniques used in their production, result in experimental values of mobility much lower than the predicted ones.

Recently, the mobility in Si–Ge strained heterostructures (Section 5.3.2) has been the object of several investigations in an effort to fabricate high frequency bipolar heterostructure transistors (Section 9.3) and high electron mobility transistors based on silicon technology. As we know, holes in III-V compounds and heterostructures have very high effective masses. Therefore, 2D hole transport in Si–Ge heterostructures with large valence band discontinuities has been intensively investigated, proving that hole mobilities as high as $10^5 \, \mathrm{cm^2 V^{-1} s^{-1}}$ can be attained.

6.2.3. Hot electrons in parallel transport

In some kinds of field effect transistors (Section 9.5) and in some nanostructures, electrons are accelerated by the electric field to kinetic energies much higher than their energies at thermal equilibrium, which are of the order of kT. After the acceleration by high electric fields, the electron energy distribution corresponds to an effective temperature higher than that of the crystal lattice, and the electrons receive the name *hot electrons*. In this situation, the new electron energy distribution is said to be decoupled from that of the lattice. Following a semi-classical approach, the effective electron temperature T_e of the electron distribution of average energy $\overline{E}$ is defined by the equation

$$\overline{E} = \frac{3}{2} k T_e \qquad (6.1)$$

Hot electron transport has been widely studied in bulk semiconductors and since the 1980s in nanostructures. Studies of hot electron parallel transport in AlGaAs/GaAs heterostructures have shown that the electron velocities reached under the action of an electric field are higher than in bulk GaAs and that the difference becomes larger at low temperatures (Figure 6.4) [4]. The increase in velocity has been attributed to the quantization of electron energies in quantum wells. The value of the velocity is specially high for the lowest subband ($E = E_1$) in comparison to the second subband ($E = E_2$) for which the electron wave function extends much more outside the barrier region and, as a consequence, the carriers are located closer to the charged donors, thus increasing impurity scattering.

An interesting effect, called *real-space transfer (RST)*, arises for hot electron parallel transport in quantum heterostructures and constitutes the basis of a new kind of high frequency devices. If the energy of the hot electrons is high enough, some of them will be able to escape from the well as indicated in Figure 6.5 [5] for an AlGaAs/GaAs/AlGaAs quantum well, for which the electrons are transferred in real space from the undoped GaAs to the surrounding AlGaAs doped semiconductor. In a low-dimensional electronic device as the one shown in Figure 6.5(b), electrons can be transferred from a high electron mobility material (GaAs) to one with a lower mobility (AlGaAs) as the voltage between source and drain is increased. As a consequence, a *negative differential resistance (NDR)* region in the *I–V* characteristics (Figure 6.5(c)) is observed. As will be shown in Chapter 9, the NDR effect leads to new kinds of devices such as resonant tunnelling transistors.

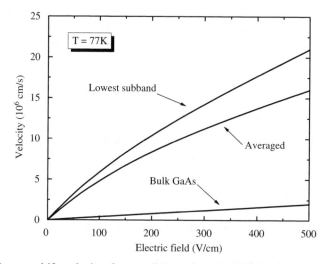

Figure 6.4. Electron drift velocity for parallel motion in AlGaAs–GaAs modulation-doped heterostructures (the value for bulk GaAs is also shown). After [4].

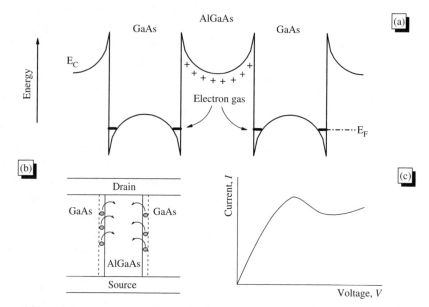

Figure 6.5. (a) Schematics of the RST mechanism; (b) structure of a device based on RST; (c) *I–V* characteristics. After [5].

In bulk samples, transport under the action of electric fields is studied in regions of dimensions much larger than the electron mean free path. However, in modern field effect devices (e.g. the MODFET) the source–drain distance and the gate lengths are very short, of the order of hundreds of nanometres. Therefore, electrons can be accelerated under the action of an electric field without suffering any collision. These electrons are called *ballistic electrons* and can reach drift velocities of the order of 10^7 cms^{-1}, higher than the saturation drift velocities by a factor as much as two. This is called the *velocity-overshoot effect*, which can be used in FET's to make the transit time of electrons between source and drain shorter, and consequently operate at higher frequencies (Chapter 9).

6.3. PERPENDICULAR TRANSPORT

In this section, we study the motion of the carriers perpendicularly to the planes of the potential barriers separating quantum heterostructures. This kind of transport is often associated to quantum transmission or tunnelling, since the carriers do not need to have enough energy to surmount the barriers. When a particle goes through a potential barrier, the wave function and its derivative (in the perpendicular direction) must be continuous, which leads to transmitted and reflected wave functions. As we will see, tunnelling through potential barriers will also lead us to the concept of negative differential resistance in the *I–V* characteristics, a phenomenon already observed by Esaki in 1957. It is also interesting that sixteen years later at IBM, Esaki, together with Tsu, proposed the observation of negative differential resistivity effects across AlGaAs/GaAs superlattices, arising from resonant tunnelling (RT) through the barriers. However, it was not until the beginning of the 1980s that heterojunctions of enough quality could be fabricated and used in RT diodes and transistors (Chapter 9).

6.3.1. *Resonant tunnelling*

Resonant tunnelling (RT) through a potential double barrier is one of the quantum vertical transport effects in nanostructures with more applications in high frequency electronic diodes and transistors as we will see in Chapter 9. Figure 6.6(a) shows the energy band diagram of a double barrier nanostructure made of undoped GaAs surrounded by AlGaAs in each side, and Figures 6.6(b) and 6.6(c) show the same structure under increasing applied voltages [6]. RT occurs for a voltage $V_1 = 2 E/e$, where E coincides (Figure 6.6(b)) with the quantized energy level E_1. In this situation, the Fermi level E_F of the metallic contact on the left coincides with the $n = 1$ level in the well. Then, the tunnelling transmission coefficient approaches unity and a large current flows through the structure. As the voltage increases over $2E_1/e$ (Figure 6.6(c)), E_F surpasses E_1 and the current through the structure decreases. Figure 6.6(d) shows the variation of the current as a

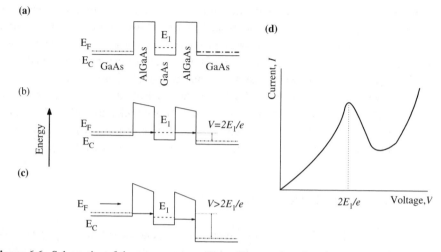

Figure 6.6. Schematics of the resonant tunnelling effect, as described in the text. After [6].

function of voltage. Evidently, for high values of V, the barriers that electrons have to tunnel become much smaller and the current increases again. This qualitative explanation was quantitatively described by Tsu and Esaki, both in diodes and superlattices quantum heterostructures grown by molecular beam epitaxy. The most important feature in the I–V characteristics of Figure 6.6(d) is that after the maximum, the slope of the curve becomes negative, i.e. there exists a region with differential negative resistance. To study the transmission properties of the double barrier, use is made of the results of a single barrier. The transmission probability in a single barrier $T(E)$ continually increases with the energy E of the incident electrons if $E/e < V_0$. The case of a double barrier is completely different and $T(E)$ is given by the product of T_E for the first barrier or emitter and that of the second one T_C or collector, that is,

$$T(E) = T_E T_C \tag{6.2}$$

Again we are only interested in the case where E is smaller than the barriers' height. To find $T(E)$ the so-called transfer matrix method described in quantum mechanics and optics texts is used. This method relates the coefficients of the incident and reflected wave functions of the neighbouring adjacent barriers by a 2×2 matrix, called transfer of propagating matrices. The case in which both barriers are identical is particularly simple to solve. The transmission coefficient of the structure is then given by [3]:

$$T(E) = \frac{T_0^2}{T_0^2 + 4R_0 \cos^2(ka - \theta)} \tag{6.3}$$

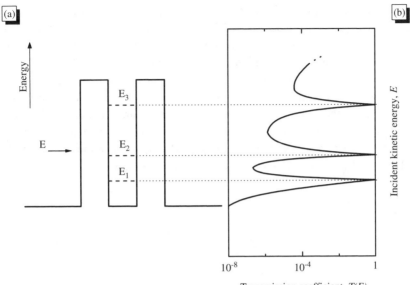

Figure 6.7. (a) Electron of energy E impinging on a resonant double barrier. The quantum well has energy levels E_1, E_2, E_3; (b) transmission coefficient as a function of incident energy.

where T_0 and R_0 are the transmission and reflection coefficients of the single barrier, a is the well thickness, k the electron wave number of the wave function in the well, and θ is the phase angle.

Figure 6.7 shows the dependence of $T(E)$ as a function of E for an RT structure with three energy levels in the quantum well. Observe that the transmission coefficient is one, at energies corresponding to the three levels, i.e. when the energy of the incident electron is aligned with these levels. The width of the resonance peaks increases with energy. This can be qualitatively explained by the Heisenberg uncertainty principle since ΔE should be inversely proportional to the lifetime τ of the states in the well. Electrons at higher levels in the well have to tunnel through lower effective barriers and should therefore have a shorter τ.

6.3.2. *Electric field effects in superlattices*

We have seen in Chapter 5 that electron states in superlattices are grouped in electronic bands or minibands, which are very narrow in comparison with bands in crystals (Figures 5.11 and 5.12). The small width of the bands and the energy gaps, or minigaps, was a consequence of the much larger dimensions of the superlattice period d in comparison with the lattice constant of a crystal, a. As we will describe in this section, electrons in narrow bands, under the action of an electric field will reveal some observable effects, such as *Bloch oscillations*, which were theoretically predicted several decades before. Also, the energy levels in each quantum well of width a of the superlattice will form a Stark ladder of step height eFa, where F is the applied electric field.

Let us suppose an electronic band in k-space such as the one shown in Figure 6.8, which is similar to the first miniband of a superlattice (Figure 5.12). Since F is applied in a given direction (let us call it z, perpendicular to the planes of the quantum wells), we can treat the problem as being one dimensional. The equation of motion for an electron in this band, under the action of an electric field is, from Eq. (2.57):

$$\hbar \frac{dk}{dt} = -eF \tag{6.4}$$

Since the electric field is constant, the solution of Eq. (6.4) for the wave number is

$$k(t) = k(0) - \frac{eF}{\hbar}t \tag{6.5}$$

According to this equation, the wave vector increases linearly with time. Suppose that the electron is initially at rest at the origin O in Figure 6.8 and that the direction of the field is opposite to k. Under this condition, the electron starts to move from O towards A

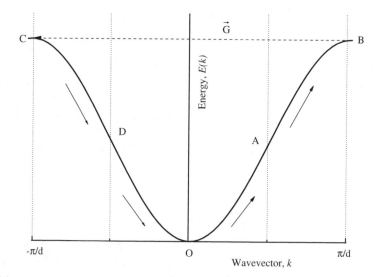

Figure 6.8. Motion in k-space of an electron in an energy band under the action of an electric field (scattering effects are neglected).

until it reaches the border of the Brillouin zone at point B ($k = \pi/d$). At B, the velocity is zero because the slope is zero (see Eq. (2.52)) and the electron is transferred to point C at $k = -\pi/d$ by the reciprocal vector $\vec{G}$, i.e. it suffers a Bragg reflection (Section 2.5.1). From C, it moves in k-space towards D by the action of the field, closing one cycle in k-space when the electron reaches O again. The motion of the electron is periodic and the velocity is given by equation:

$$v = \frac{1}{\hbar}\frac{dE}{dk} \tag{6.6}$$

also oscillates if the band energy is like the one represented in Figure 6.8. Therefore, the electron performs an oscillatory motion both in real and k-spaces.

The period T_B of the oscillatory motion in k-space is determined by the time spent in covering the width $2\pi/d$ of the Brillouin zone and given by:

$$T_B = \frac{2\pi}{\omega_B} = \frac{2\pi\hbar}{eFd} \tag{6.7}$$

Note that T_B and ω_B depend only on the periodicity of the superlattice and the electric field, but are independent of the energy width of the miniband. Evidently, in order to experimentally observe Bloch oscillations, T_B should be shorter that the relaxation time due to scattering. Bloch oscillations cannot be observed in bulk solids because their typical values of T_B ($\sim 10^{-11}$s) are much longer than the corresponding ones in a superlattice, since d can be as much as two orders of magnitude larger than the lattice constant. We can reach the same conclusion if we consider that the width of the energy bands in solids are much greater than in a superlattice. Therefore the electrons close to O in Figure 6.8, are not able to surmount the energy to values close to B at the Brillouin zone ($k = \pi/d$) since the wave number given by Eq. (6.5) cannot become large enough, due to scattering events which send back the electrons again close to O. In practice, the value of T_B cannot be made very low by making the values of F very high since this would produce Zenner tunnelling, that is, electrons originally in one tilted miniband (see Figure 6.9(b)) could tunnel across the gap to a neighbouring miniband and Bloch oscillations would not be produced. Therefore, in order to observe Bloch oscillations the minibands should be quite narrow, contrary to the minigaps.

Figure 6.9(a) shows the miniband energy structure of a superlattice in which for simplicity only two minibands are shown. If a constant electric field F is applied in the z-direction, the bands become tilted with a slope equal to $-eF$, since the expression of the potential energy becomes, in relation to the energy E_0 at the origin:

$$E(z) = E_0 - eFz \tag{6.8}$$

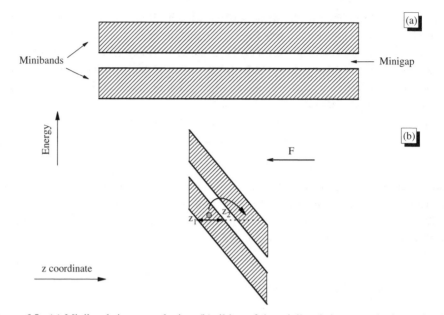

Figure 6.9. (a) Minibands in a superlattice; (b) tilting of the minibands in a superlattice under the
action of an applied electric field.

Because of the inclination of the bands, shown in Figure 6.9(b), an electron with total
energy E_T will oscillate in space between locations z_1 and z_2. If F increases, the bands
become steeper and the electron is spatially localized in a smaller region. Evidently, for
fields high enough, the electron would eventually become localized within a quantum
well. The condition for this to occur is that the energy drop ΔE between levels in two
successive walls should be higher than the energy width Δ of the minibands, since for
$\Delta E = eFd > \Delta$, the wells become completely decoupled (Figure 6.10(a)). Therefore,
for fields larger than Δ/ed, the electrons are localized in quantum wells whose eigenstates
differ considerably in energy and the concept of energy miniband breaks down. Instead,
the quantized energy states form what is known as a *Stark ladder*. Stark localization in
AlGaAs–GaAs superlattices, first observed by Méndez [6], has important applications in
electro-optical devices (Chapter 10).

Similarly to resonant tunnelling diodes, superlattices also show negative differential
resistance (NDR) regions in their *I–V* characteristics, which can be used in a series of
electron devices. The regions of NDR are observed when high electric fields are applied
through the structure and the successive quantum wells differ in energy by about eFd.
From Section 6.3.1 we know that resonant tunnelling occurs when

$$E_2 - E_1 = eFd \tag{6.9}$$

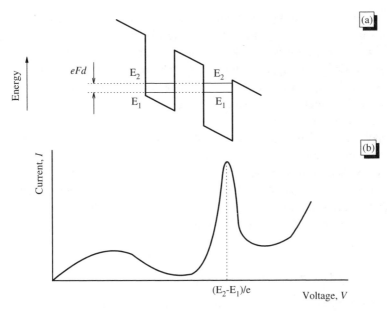

Figure 6.10. (a) For high electric fields, the superlattice miniband spectrum is destroyed and the system can be considered as a MQW differing in energy by eFd; (b) I–V characteristics of a superlattice.

where E_1 and E_2 are now the levels when F is applied, which do not have necessarily to coincide with the levels when $F = 0$. As shown in Figure 6.10(b), a region of pronounced NDR appears in the I–V curve, right after the resonance peak. Therefore superlattices, like resonant tunnelling diodes, can also be used in high-frequency oscillators and amplifiers.

6.4. QUANTUM TRANSPORT IN NANOSTRUCTURES

Next we are going to deal with *quantum transport*, which is produced when nanostructures are connected to an external current by means of leads or contacts. This transport is also called mesoscopic transport. As we explained in Section 1.3, the term "mesoscopic" refers to systems with a range of sizes between the macroscopic world and the microscopic or atomic one, and which have to be explained by quantum mechanics. These systems in electronics are also known as submicron or nanoscale devices. One very interesting phenomenon which appears in mesoscopic transport is the quantization of the conductance in units of $2e^2/h$ (Sections 6.4.1 and 6.4.2). Another very interesting phenomenon called Coulomb blockade can be observed in very small nanometric structures, like a quantum dot (Section 6.4.3).

   In order to observe quantum transport effects in semiconductor nanostructures, some conditions must be met. In general we can say that at a given temperature, quantum transport will be more easily revealed in nanostructures in which the electron effective mass is small, since this implies high electron mobilities. In addition, as was remarked in Section 4.2, the energy of electron levels in a quantum well increases when the effective mass decreases. Therefore, the smaller the effective mass, the higher the temperature at which quantum transport can be observed.

   Transport in mesoscopic devices is usually ballistic (Section 6.2.3), since the dimensions of the devices are smaller that the mean free path of electrons, which in the case of AlGaAs/GaAs heterostructures at low temperatures can be longer than several microns. Another important property of *ballistic transport*, in addition to the non-scattering properties, is that electrons do not lose their phase coherence, since they do not suffer inelastic collisions. Therefore electrons can show phase interference effects in mesoscopic systems.

6.4.1. Quantized conductance. Landauer formula

For an elementary description of quantum conductance effects, it is more appropriate to deal with 1D mesoscopic semiconductor structures like quantum wires. If the wire is short enough, i.e. shorter than the electron mean free path in the material, there would be no scattering and the transport is ballistic. Suppose, as in Figure 6.11, that the 1D quantum wire is connected through ideal leads, which do not produce scattering events, to reservoirs characterized by Fermi levels E_{F1} and E_{F2}. Suppose also that in order for the current to flow through the quantum wire, a small voltage V is applied between the reservoirs. As a consequence, there is a potential energy eV between the two reservoirs equal to $E_{F1} - E_{F2}$. The current across the wire should be given by the product of the concentration of electrons (obtained from the density of states function $n_{1D}(E)$, in the

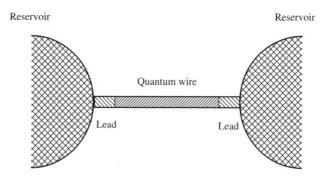

Figure 6.11. Schematics of a 1D mesoscopic system used to derive the Landauer formula.

energy interval eV), the electron velocity $v(E)$, and the unit electronic charge:

$$I = en_{1D}(E)v(E)eV \qquad (6.10)$$

Substituting $n_{1D}(E)$ by its expression given by Eq. (4.21) (without the two factor because now the electrons move in one direction), one obtains the following value of the current:

$$I = \frac{2e^2}{h}V \qquad (6.11)$$

which, interestingly enough, is independent of the carrier velocity. The value of the conductance $G \equiv (I/V)$ is therefore:

$$G = \frac{2e^2}{h} \qquad (6.12)$$

It is interesting to observe that the conductance of the quantum wire is length independent, in contrast to the classical case where it varies inversely to the length.

The quotient

$$G_0 = \frac{e^2}{h} \qquad (6.13)$$

is called the *quantum unit of conductance* and corresponds to a *quantum resistance* of value

$$R_0 = \frac{h}{e^2} = 25.812807 \, \mathrm{k\Omega} \qquad (6.14)$$

which can be experimentally determined. Since the quantity $2e^2/h$ appears very often, it is usually called *fundamental conductance*.

The above results on quantum conductance and resistance have been derived in the simplest possible manner, using a 1D mesoscopic system. This quantification of macroscopic classical concepts, like conductance and resistance, is of fundamental importance in mesoscopic physics. Before we go into deeper subjects, it is convenient to generalize the above results. One generalization which will be treated in the next section consists of the study of nanostructures with many leads, instead of the two characteristic of a 1D system. A second generalization is related to the existence of energy subbands in low-dimensional semiconductors (see, for instance, Section 4.2). Higher subbands than those corresponding to the first quantization level can participate in transport if the electron concentration or the energy becomes high enough.

In the case of quantum wires, the subbands – or channels in the language of quantum transport – arise from the transverse states (Section 4.5). Assuming the existence of several channels, let us suppose that the leads can inject electrons in any channel or mode m and that, after interacting with a scattering centre in the mesoscopic structure, the electrons emerge through any channel n. These electrons will make a contribution to the system's total conductance equal to the product of the conductance quantum $2e^2/h$ and the quantum-mechanical transmission probability $|t_{nm}|^2$ for electrons being injected through channel m and emerging through channel n. (Observe that, in this formulation, the transmission probability is expressed in terms of the amplitude or transmission coefficients t_{nm} of the electron wave functions.) Therefore, the total conductance will be obtained by adding over all channels:

$$G = \frac{2e^2}{h} \sum_{n,m}^{N} |t_{nm}|^2 \qquad\qquad (6.15)$$

where N is the number of quantum channels contributing to the conductance. Eq. (6.15) can be considered as a generalization of Eq. (6.12) for a mesoscopic system with two leads and many channels, and is called the *Landauer formula*.

For many quantum transport studies, a nanostructure consisting of a small constriction within a 2D system is used. Such a structure is illustrated in the inset of Figure 6.12 and is based on a split gate acting on the electrons of a 2D heterostructure. Because of the special shape of the gate, the electrons in the 2D plane are constrained to travel through a very small or quasi-1D region, as a consequence of the distribution in electrical voltage when an external voltage is applied to the gate. This structure is called *quantum point contact (QPC)* or an *electron waveguide*, because of the analogy of the structure with waveguides in electromagnetism.

Figure 6.12 shows the first observation of quantum conductance by Wees et al. in 1988 in the QPC structure (see inset) formed in an AlGaAs/GaAs quantum heterostructure [7]. As can be appreciated, the values of the conductance are quantized in multiples of the fundamental conductance $2e^2/h$ when the gate voltage is varied. This quantification can be shown to arise from Eq. (6.15) since the transmission probability coefficients approach unity for very low scattering rates as is the case in the QPC. The observation of sharp plateaus in the *I–V* characteristics is not always easy since the sharp *I–V* structure can be degraded by several factors: inelastic scattering effects, non-zero resistance of the electrical contacts to the leads, impurities, surface roughness, etc. As a consequence, the magnitudes of the experimental values of the conductance steps can vary by a few percent, as shown in the figure. In contrast, we will see in the next chapter that, under the effect of strong magnetic fields, the accuracy in the conductance steps can be about 10^6 times better. This is the reason why the quantum Hall effect (Section 7.7) finds many applications in metrology.

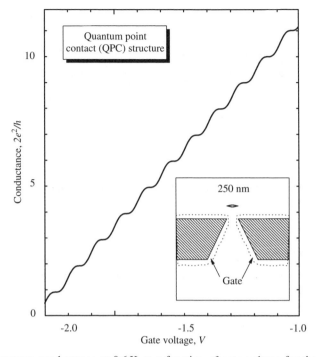

Figure 6.12. Quantum conductance at 0.6 K as a function of gate voltage for the quantum point
contact, shown in the inset, created within an AlGaAs–GaAs heterojunction. After [7].

6.4.2. *Landauer–Büttiker formula for multi-probe quantum transport*

In the preceding section we have derived Eq. (6.15) for quantum transport in a two-lead
nanostructure. Now let us generalize this equation to the case of nanostructures with
several leads. Let us suppose a nanostructure like the one shown in Figure 6.13, often
used in quantum Hall effect measurements (Section 7.7), with two current leads connected
to corresponding reservoirs and several voltage probes. The reservoirs serve as an infinite
source and sink of electrons and are kept at constant temperature, even if they provide
or take electrons from the structure. We will calculate as before the current in any lead
i connected to reservoir μ_i by assuming first that there is only one channel in each lead.
Similarly, we also follow the procedure of the scattering or transmission matrix formed by
the transmission coefficients T_{ij} connecting leads i and j. Since electrons incident to the
structure from any lead can be reflected, we also make use of the reflection coefficients R_i.
To find the current I_i in lead i, we have to take into account the following contributions:
(i) the current injected through the lead taken from reservoir μ_i which is given by $(2e/h)\mu_i$;

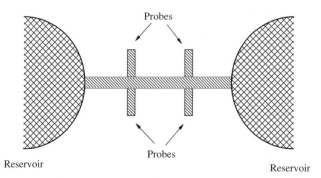

Figure 6.13. Diagram of a typical nanostructure used to make quantum Hall effect measurements (see text).

(ii) the fraction R_i of this current which is reflected back into the lead, and (iii) all the currents from the other leads injected into lead i. The sum of all these contributions, with their respective signs, gives the current I_i in lead i:

$$I_i = \frac{2e^2}{h}\left[(1 - R_i)\, V_i - \sum_{j \neq i} T_{ij} V_j\right] \tag{6.16}$$

where use has been made of the voltages V_i corresponding to μ_i, i.e. $\mu_i = eV_i$. We should point out that the values of V_i in the above expression are referred to a common voltage $V_0 = \mu_0/e$, where μ_0 corresponds to the lowest energy level of the Fermi distribution in the reservoirs, below which all energy levels are occupied and therefore cannot contribute to the current. Evidently, at temperatures close to $T = 0\,$K, μ_0 should coincide with the Fermi value of the smallest value of all μ_i.

The above equation was obtained for leads with one single channel. The multiprobe generalization assumes that in each lead i, there are N_i propagating channels. We define generalized transmission coefficients $T_{ij,\alpha\beta}$ for the probability of a carrier in lead j and channel β to be transmitted into lead i and channel α. In the same manner, generalized transmission coefficients $R_{i,\alpha\beta}$ are also defined for the probability of a carrier being reflected from channel β into channel α, both in lead i. Proceeding now similarly to the case of one channel, we add the contributions of all possible currents to the total current in lead i and obtain:

$$I_i = \frac{2e^2}{h}\left[(N_i - R_i)\, V_i - \sum_{j \neq i} T_{ij} V_j\right] \tag{6.17}$$

where V_i is the voltage of reservoir i and T_{ij} and R_i are the reduced transmission and reflection coefficients defined by

$$T_{ij} = \sum_{\alpha,\beta} T_{ij,\alpha\beta} \text{ and } R_i = \sum_{\alpha,\beta} R_{i,\alpha\beta} \tag{6.18}$$

Eq. (6.17) is called the *Landauer–Büttiker formula* for multi-probe quantum transport for leads with several channels. As in the equation for the quantized conductance of Section 6.4.1, the fundamental conductance factor, $2e^2/h$, appears again in the expression for the current.

We can further simplify this equation if we take into account the conservation of current in the mesoscopic structure. The net current I_i injected into the structure by lead i, which equals $(2e^2/h)N_iV_i$ minus the reflected fraction $(2e^2/h)R_iV_i$, should have the same value as the sum of all currents that originating from I_i leave the structure, i.e. $(2e^2/h)\sum_{j\neq i} T_{ij}V_{ij}$

Therefore, we should have

$$N_i - R_i = \sum_{j\neq i} T_{ij} \tag{6.19}$$

Using this result, Eq. (6.17) can also be written in the form

$$I_i = \frac{2e^2}{h} \sum_{j\neq i} T_{ij}(V_i - V_j) \tag{6.20}$$

which is another expression of the Landauer–Büttiker formula.

6.4.3. Coulomb blockade

We know that in microelectronic devices like the MOSFETs the magnitude of the currents are reduced as the feature size of the device shrinks. In the limit we can ask ourselves what happens when the current is transported by just one single electron. Imagine a semiconductor of nanometric size in the three spatial dimensions, such as for instance, a quantum dot (Section 4.6). We will show in this section that even the change of one elementary charge in such small systems has a measurable effect in the electrical and transport properties of the dot. This phenomenon is known as *Coulomb blockade*, which we will discuss in the simplest possible terms.

Let us imagine a semiconductor quantum dot structure, connected to electron reservoirs at each side by potential barriers or tunnel junctions (Figure 6.14(a)). In order to allow the transport of electrons to or from the reservoirs, the barriers will have to be sufficiently

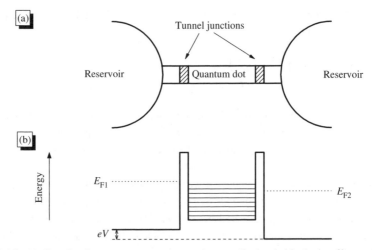

Figure 6.14. (a) Sketch of a quantum system to observe Coulomb blockade effects; (b) energy
distribution when an external voltage is applied through the dot. The levels shown in
the dot depend on the number N of electrons in it.

thin, so that the electrons can cross them by the tunnel effect. Figure 6.14(b) represents
the energies of the quantum dot when the number of electrons in the dot, N, is changed
in amounts of single unit charges. Evidently potential differences can be provided if an
external voltage source is connected to them.

Suppose that we wish to change the number N of electrons in the dot by adding just
one electron, which will have to tunnel for instance from the left reservoir into the dot.
For this to happen, we will have to provide the potential energy eV to the electron by
means of a voltage source. If the charge in the quantum dot is Q and its capacitance C,
the potential energy is $Q^2/2C$. Therefore an energy of at least $e^2/2C$ will have to be
provided to the electron, which means that for the electron to enter the dot, the voltage
will have to be raised to at least $e/2C$. Since the electron can either enter the dot or leave
it (this process is equivalent to a hole entering the dot), we see that electrons cannot
tunnel if

$$|V| < \frac{e}{2C} \tag{6.21}$$

Therefore, there is a voltage range, between $-e/2C$ and $e/2C$, represented in Figure 6.15
in which current cannot go through the dot, hence the name of Coulomb blockade given
to this phenomenon.

Evidently if the above process is continued and we keep adding more electrons, we will have the situation represented in Figure 6.16, in which we will observe discontinuities in the current through the quantum dot whenever the voltage acquires the values expressed by:

$$V = \left(\frac{1}{2C}\right)(2n+1)e, \qquad n = 0, 1, 2, \ldots \qquad (6.22)$$

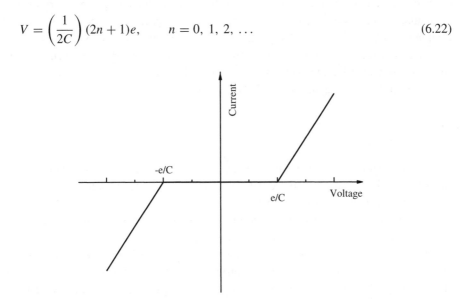

Figure 6.15. *I–V* characteristics in a quantum dot showing the Coulomb blockade effect.

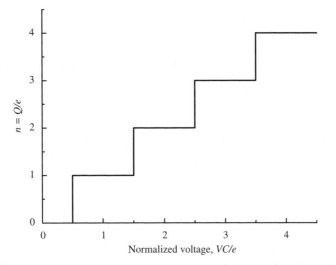

Figure 6.16. Charging of a quantum dot capacitor as a function of voltage, in normalized coordinates (see text).

Observe that in Figure 6.16 we have made use of normalized coordinates, both in the horizontal and vertical axes, to better appreciate the effect of the quantification in current and voltage.

It is also interesting to observe from the above equations that as the size of the quantum dot is reduced, and therefore C gets smaller, the value of the energy necessary to change the number of electrons in the dot increases. In this case, it will be easier to observe the Coulomb blockade, since the changes in voltage and electric energy for electrons to enter the dot also increase. This change in electric energy has to be much larger than the thermal energy kT at the working temperature, in order to observe measurable Coulomb blockade effects. Therefore, we should have for the capacitance

$$C \ll \frac{e^2}{kT} \tag{6.23}$$

For this condition to be fulfilled, either the capacitance of the dot should be very small (values less than 10^{-16} F are very difficult to get) or we should work at very low temperatures, usually smaller than 1 K.

Another condition to observe single electron effects is that the number of electrons in the dot should not fluctuate in equilibrium. Let us assume that the time taken for an electron to be transferred in or out of dot is of the order of $R_T C$, where R_T is the equivalent resistance of the tunnel barrier and C the capacitance of the dot. Fluctuations in the number of electrons in the dot induce changes in potential energy of the order of e^2/C. Therefore we should have, according to the uncertainty principle,

$$\Delta E \, \Delta t = \frac{e^2}{C} R_T C > h \tag{6.24}$$

and consequently for Coulomb blockade effects to be clearly observed we should have

$$R_T \gg \frac{h}{e^2} = 25.8 \, \text{k}\Omega \tag{6.25}$$

In single electron transport experiments, usually the current is measured, which is proportional to the conductance G. In terms of the conductance, the above condition can be written as

$$G \ll \frac{e^2}{h} \tag{6.26}$$

A very interesting challenge of future electronics is the control of switching devices by just one electron. In Section 9.7, we will consider the single electron transistor (SET) whose functioning is based on single electron effects.

REFERENCES

[1] Davies, J.H. (1998) *The Physics of Low-Dimensional Semiconductors* (Cambridge University Press, Cambridge).
[2] Störmer, H.L., Gossard, A.C. & Wiegmann, W. (1982) *Solid State Commun.*, **41**, 707.
[3] Masaki, K., Taniguchi, K., Hamaguchi, C. & Iwase, M. (1991) *Jpn. J. Appl. Phys.*, **30**, 2734.
[4] Schubert, E.F. & Ploog, K. (1984) *Appl. Phys.*, **A33**, 183.
[5] Luryi, S. & Zaslasvsky, A. (1998) In *Modern Semiconductor Device Physics*, Ed. Sze, S.M. (Wiley, New York).
[6] Méndez, E.E. (1987) In *Physics and Applications of Quantum Well and Superlattices*, Eds. Méndez, E.E. and von Klitzing, K., NATO ASI Series B, Physics, **170**, 159–188 (Plenum, New York).
[7] van Wees, B.J. (1988) *Phys. Rev. Lett.*, **60**, 848.

FURTHER READING

Ferry, D.K. & Goodnick, S.M. (1997) *Transport in Nanostructures* (Cambridge University Press, Cambridge).
Hamaguchi, Ch. (2001) *Basic Semiconductor Physics* (Springer, Berlin).
Mitin, V.V., Kochelap, V.A. & Stroscio, M.A. (1999) *Quantum Heterostructures* (Cambridge University Press, Cambridge).
Schäfer, W. & Wegener, M. (2002) *Semiconductor Optics and Transport Phenomena* (Springer, Berlin).

PROBLEMS

1. **Impurity scattering for parallel transport**. The mobility of electrons for parallel transport at low temperatures in a 2D structure similar to that of Figure 5.4 is limited by ionized impurity scattering and is approximately given by (see Ref. [1])

$$\mu \approx 16|d|^3 \frac{e}{\hbar} (2\pi n_{2D})^{1/2}$$

where d is the distance from the plane where electrons move to the plane in which the impurities are located and n_{2D} is the density of ionized impurities. (a) Calculate the values of μ for the cases $d = 10\,\text{nm}$ and $d = 20\,\text{nm}$, and $n = 10^{16}\,\text{m}^{-2}$. Compare the calculated values of μ to those of Figure 6.2 at low temperatures. (b) Explain qualitatively the strong dependence of μ on d, as well as the compromise that should be reached between large and small values of d.

2. **Resistance of one channel**. (a) From the Landauer formula for the total conductance of a system, Eq. (6.15), show that the resistance for the case of just one channel can be expressed as

$$R = \frac{h}{2e^2} + \frac{h}{2e^2}\frac{R}{T}$$

where R and T are the transmission and reflection coefficients. (b) Show that if the channel of the above example is formed by a material which is a perfect conductor, then

$$R = \frac{h}{2e^2}$$

Explain why the above value for the resistance is different from zero.

3. **Voltage probes**. Suppose a system formed by one 1D conductor between two contacts, 1 and 2, which transports a current I when a difference of voltage V is applied between them. A voltage probe (contact 3) is connected in the middle between contacts 1 and 2. Show that the voltage V_3 at this probe is given by $V/2$, i.e. the same value as we would expect classically. *Hint*: for the voltage probe 3 the current is $I = 0$. With this value for I, obtain from the Landauer–Büttiker formula, Eq. (6.17), the expression for V_3 and assume that by symmetry considerations $t_{3,1} = t_{3,2}$.

4. **Incoherent transmission through a double barrier**. Suppose two tunnelling barriers in series separated by a small distance. Show that the transmission T through the double barrier is given by

$$T = \frac{|t_1|^2|t_2|^2}{1 - |r_1|^2|r_2|^2}$$

where t_1, t_2 are the transmission amplitudes of each barrier and r_1, r_2 the corresponding reflection amplitudes. *Hint*: if you call α the amplitude from barrier 1 inciding in barrier 2, β the amplitude from barrier 2 inciding in barrier 1, and γ the outgoing amplitude from barrier 2, we have

$$|\alpha|^2 = |t_1|^2 + |r_1|^2\,|\beta|^2$$
$$|\beta|^2 = |\alpha|^2\,|r_2|^2$$
$$|\gamma|^2 = |\alpha|^2\,|t_2|^2$$

5. **Composition of two series resistances**. Suppose two tunnel barriers in series with the amplitudes of transmission and reflection given in the previous problem. Show that the total resistance is equal to

$$R = \frac{h}{2e^2}\left(1 + \frac{|r_1|^2}{|t_1|^2} + \frac{|r_2|^2}{|t_2|^2}\right)$$

Compare with the classical result derived from Ohm's law.

6. **Two-terminal and four-terminal resistances**. A current of value I is passed between contacts 1 and 2 along the two ends of a longitudinal structure. Voltage probes 3 and 4 are placed between contacts 1 and 2. Define a two-terminal resistance $R_{12,12}$ as the quotient of the voltage V_{12} between contacts 1 and 2 divided by the current I_{12} between these contacts. We also define a four-terminal resistance $R_{12,34}$ as the quotient of the difference of potential between the voltage probes 3 and 4 divided by I_{12}. By following the procedure of Landauer–Büttiker show that:

$$R_{12,12} = \frac{h}{2e^2}\frac{1}{t_{12}}$$

$$R_{12,34} = \frac{h}{2e^2}\frac{1}{t_{12}}\frac{t_{31}t_{42} - t_{32}t_{41}}{(t_{31} + t_{32})(t_{41} + t_{42})}$$

(Perhaps, at this time you wonder why both results are different, since the voltage probes take no current. This point has arisen a long controversy.)

7. **Observation of single electron effects**. Suppose a metallic quantum dot, of shape similar to a flat circular disk of radius R parallel to an infinite metal plane, at a distance L from the plane. Show that in order to observe single electron effects at room temperature, the radius of the dot should be of the order of a few nanometers. Explain also the situation if the temperature is very low. *Hint*: assume that we know from electrostatics that if $R \ll L$ the capacitance of the dot is $C = 8\varepsilon_0\varepsilon_r R$. Take as value of ε_r the relative dielectric constant of silicon.

8. **Single-electron turnstile**. The single-electron turnstile consists of a quantum dot that, under the action of an alternating voltage, transfers one electron through the dot each cycle. Two AC signals, 180 degrees out of phase, are applied to each barrier surrounding the dot, thus shifting the heights of the wells by equal and opposite amounts. Show the operation of the turnstile considering the quantum dot structure of Figure 6.14(a) and that a constant current $I = ef$ is produced by the turnstile where e is the electron charge and f the frequency of the applied AC signal.

Chapter 7

Transport in Magnetic Fields and the Quantum Hall Effect

Chapter 7

Transport in Magnetic Fields and the Quantum Hall Effect

7.1. INTRODUCTION

This chapter is mainly focused on the study of transport properties of the 2D electron systems in magnetic fields, which led to the discovery of the quantum Hall effect, one of the most monumental findings in modern solid state physics. The significance of the quantum Hall effect is reflected in the fact that Nobel prizes were awarded to von Klitzing in 1985 and Tsui, Stormer, and Laughlin in 1999, for the discovery of the integral and fractional quantum Hall effects, respectively. The chapter starts, Section 7.2, with a brief review of the effects of magnetic fields on bulk metals and semiconductors, in particular the quantization of the orbital motion of the electrons into Landau levels. The next Sections 7.3 and 7.4 deal with the action of magnetic fields on 2D electron systems formed at quantum heterojunctions. In this case, the constant 2D density of states (DOS) function collapses into δ-functions (in real systems broadened by electron scattering), as a consequence of the degeneracy of the Landau levels.

The effects of magnetic fields on the 2D electron systems have important practical consequences. In Section 7.5, we show that currents in mesoscopic systems can produce interference effects which are manifested by the modulation of the conductance under the action of a magnetic vector potential $\vec{A}$ (Aharonov–Bohm effect). Another very interesting effect is demonstrated by the Shubnikov–de Haas oscillations (Section 7.6), which are observed in the conductance of the 2D systems under the action of a magnetic field, and are much more pronounced than in bulk semiconductors.

With this background, the reader is now prepared to face in Section 7.7 the quantum Hall effects. According to the integral quantum Hall effect (IQHE, Section 7.7.1), the Hall voltage of a mesoscopic 2D electron system is quantized in voltage "plateaux". Consequently, both, the conductance and its inverse resistance, are also quantized. The impressive fact about this quantification is that it is both independent of the semiconductor materials forming the quantum heterostructure, and of the geometry of the sample. The values of the voltage "plateaux" are so precise that they have led to the introduction of a new universal constant, the von Klitzing constant ($R_K = h/e^2$), which can be measured with a precision of a few parts in 10^{10}. This resistance has been used since 1990 for the determination of the resistance standard (ohm) in the SI system. In Section 7.7.2, the link between the formalism of the IQHE and the Landauer–Büttiker formulation of quantum transport, originally introduced in Chapter 6, is established through the new concepts of skipping orbits and edge states. The physics of the IQHE culminates (Section 7.7.3) with the introduction of the concepts of localized states for explaining the width of the Hall "plateaux".

The experimental results of the fractional quantum Hall effect (FQHE) might appear somewhat similar to those of the IQHE, although its theoretical explanation is completely different. The understanding of the FQHE is based on many-body interactions between electrons at very low temperatures under the action of magnetic fields. Although a theory explaining all aspects of the FQHE does not exist yet, we present in Section 7.7.5 some concepts of the two theories that in our opinion have the most credit. The theory of Laughlin (1983) is based on many-body electron interactions, which explains the fractional values of the Landau levels filling factor, and even postulates excitations of quasi-particles with a fraction (e.g. 1/3) of the electron charge. The theory of the composite fermion (CF), introduced by Jain (1989, 1990), postulates the CF as a new entity consisting of one electron with two attached fluxons $\Phi_0(= e/h)$. This theory, which predicts the existence of quasi-particle excitations of a charge equal to a fraction of the electron charge has recently been receiving much support.

7.2. EFFECT OF A MAGNETIC FIELD ON A CRYSTAL

First let us review the effect of a magnetic field on the conduction electrons in a solid. We know from solid state physics that the application of high magnetic fields to a crystal has remarkable effects, among them, the collapse of energy states of conduction electrons into Landau levels, oscillations in the magnetization $\vec{M}$ when the magnetic flux density $\vec{B}$ varies (de Haas–van Alphen effect), oscillations of the electrical resistivity as a function of the magnetic field (Shubnikov–de Haas effect), etc. These effects are due to the quantization of the energy of the conduction electrons under the action of a magnetic field B_z, applied in the z-direction, according to:

$$E_n = \left(n + \frac{1}{2}\right)\hbar\omega_c + \frac{\hbar^2 k_z^2}{2m_e^*}, \quad n = 0, 1, 2, \ldots \tag{7.1}$$

where ω_c is the cyclotron frequency given by

$$\omega_c = \frac{eB_z}{m_e^*} \tag{7.2}$$

which corresponds to the frequency of the cyclotron orbits that the electrons perform in the (x, y) plane. In Eq. (7.1) the quantum number n corresponds to the different

Landau levels. On the other hand, the field does not alter the motion of the electrons along the z-direction. As shown in Eq. (7.1) and Figure 7.1(a) the electrons behave with respect to the z-direction as if they were free. This behaviour is expected because a magnetic field cannot exert a force to an electron moving parallel to it. On the other hand, the electron motion in the x and y directions is quantized according to a quantum harmonic oscillator model.

Let us now discuss the effect of the previous magnetic field on the 3D density of states. As we know from solid state physics, when a magnetic field B_z is applied, the 3D allowed states in k-space collapse into a set of concentric tubes parallel to $\vec{B}$, which leads to the result that each Landau tube presents a degeneracy given by $g_n = eB/\pi\hbar$. In addition, since each Landau level is associated with a 1D free electron behaviour along the direction of $\vec{B}$, the energy dependence of the (DOS) function n_{2D} should be of the form (see Eq. (4.20)), $g(E) \propto 1/\sqrt{E}$. The DOS function should also present singularities (Figure 7.1(b)) at the bottom of each subband, corresponding to the respective Landau level. Evidently, in a practical situation, the singularities which appear at every Landau level ($n = 1, 2, 3, \ldots$) are removed as a consequence of electron scattering.

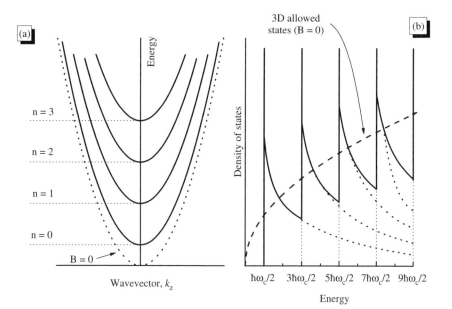

Figure 7.1. (a) Electron energy bands for a 3D solid vs the z-direction wave vector for different Landau levels ($n = 0, 1, 2\ldots$); (b) density of states function for the Landau levels compared with the free-electron gas for the case $B = 0$.

7.3.   LOW-DIMENSIONAL SYSTEMS IN MAGNETIC FIELDS

Unlike the case for bulk crystals (3D), where, under the action of $\vec{B}$, the electron system was quantized in the plane perpendicular to $\vec{B}$, in a 2D electron system, the energy spectrum becomes completely quantized. In order to show this, let us proceed to write down Schrödinger's equation for an electron in a 2D system under the action of a magnetic field applied in a direction (z) perpendicular to the low-dimensional system. In this derivation we make use of the Landau gauge, in which the vector potential $\vec{A}$ has only one component, let us say A_y, such that $A_y = Bx$. As we recall from any text on quantum mechanics, Schrödinger's equation in an electromagnetic field is obtained from its regular expression by substituting the canonical momentum $\vec{p}$ by $\vec{p} - q\vec{A}$ (Peierl's substitution). Therefore the expression of Schrödinger's equation becomes for the wave function $\psi(\vec{r}) = \psi(x, y)$ of the 2D system:

$$\left[-\frac{\hbar^2}{2m}\frac{\partial^2}{\partial x^2} + \frac{1}{2m}\left(i\hbar\frac{\partial}{\partial y} + eBx \right)^2 \right]\psi(x, y) = E\psi(x, y) \tag{7.3}$$

Operating in this equation:

$$\left[-\frac{\hbar^2}{2m}\left(\frac{\partial^2}{\partial x^2} + \frac{\partial^2}{\partial y^2} \right) - \frac{i\hbar eBx}{m} + \frac{(eBx)^2}{2m} \right]\psi(x, y) = E\psi(x, y) \tag{7.4}$$

Let us now try a solution of the form:

$$\psi(x, y) = \varphi(x)e^{iky} \tag{7.5}$$

to make Schrödinger's equation separable. The plane wave corresponding to the y coordinate (free electron motion) is suggested by the fact that $\vec{A}$ does not depend on y. Substituting the expression of the wave function given by Eq. (7.5) into Eq. (7.4), the plane wave dependence cancels, leaving the following equation in x:

$$\left[-\frac{\hbar^2}{2m}\frac{d^2}{dx^2} + \frac{1}{2}m\omega_c^2(x - x_0)^2 \right]\varphi(x) = E_n\varphi(x) \tag{7.6}$$

where ω_c is given by Eq. (7.2) and

$$x_0 = \frac{\hbar k}{eB} \tag{7.7}$$

Eq. (7.6) can be recognized immediately as Schrödinger's equation for a one dimensional harmonic oscillator, since the term x_0, added to x, only implies that the origin of the parabolic potential is displaced by x_0. (For this reason x_0 is known as the coordinate centre.) We therefore reach the important conclusion that the eigenstates of the 2D system in a magnetic field are given by:

$$E_n = \left(n + \frac{1}{2}\right)\hbar\omega_c, \quad n = 0, 1, 2, 3, \dots \tag{7.8}$$

Note that the energy attributed to the magnetic field depends only on the quantum number n and the magnetic field B through ω_c.

7.4. DENSITY OF STATES OF A 2D SYSTEM IN A MAGNETIC FIELD

We have seen in the previous section that if a strong magnetic field is applied perpendicularly to a quasi-two-dimensional electron system, the electrons adopt a cyclotron orbital motion of frequency ω_c given by Eq. (7.2), and that the energy is quantified according to a 1D harmonic oscillator model. Therefore, the DOS function of the 2D gas (with $B = 0$), collapses into a δ-function at each Landau level as a consequence of the application of B.

Figure 7.2 shows how the n_{2D} DOS function, which is constant, i.e. independent of energy for each subband, as given in Section 4.2, collapses into δ-functions at each Landau

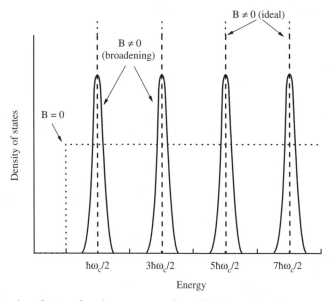

Figure 7.2. Density of states function vs energy for a 2D electron gas in a magnetic field. (For comparison, the curve corresponding to a 2D system with $B = 0$ is also shown.)

level when a magnetic field is applied in a direction perpendicular to the 2D system. As we can appreciate in Figure 7.2, the lowest Landau level is located at an energy, $\hbar\omega_c/2$, higher than the energy corresponding to the bottom of the parabolic subband. Since the electrons suffer scattering events in their motion through the crystal, the δ-function states broaden as shown in the figure. This broadening effect, caused by defects such as impurities or by lattice vibrations (phonons), is characterized by a characteristic energy width Γ. Evidently, for the Landau levels to become well identified, $\hbar\omega_c > \Gamma$. This is equivalent to saying that the scattering time in transport τ_i should be sufficiently large or that the mobility of the electron in the system should be high enough.

Since all the levels in an interval $\hbar\omega_c$ collapse into the same Landau level when the field B is applied, the *degeneracy D* of each Landau level should be given by

$$D = \frac{m_e^*}{2\pi\hbar^2}\hbar\omega_c = \frac{eB}{2\pi\hbar} \tag{7.9}$$

where use has been made of the expression for the DOS n_{2D} (Eq. (4.7) in Chapter 4), but without taking into account spin degeneracy. Note from Eq. (7.9) that the degeneracy of the Landau levels increases linearly with the magnetic field, a fact which will have important consequences in the explanation of the quantum Hall effects in Section 7.7.

7.5. THE AHARONOV–BOHM EFFECT

Magnetic fields can produce and control interference effects between the electrons in solids. Evidently, in order to observe interference effects between different electron waves, their phase has to be maintained. Recall from Section 1.4 that we defined the phase coherence length L_ϕ as the distance travelled by an electron without changing its phase. The phase of an electron wave is generally destroyed when electrons interact inelastically with defects in the lattice. In general, ballistic electrons (Section 1.8) with a mean free path ℓ much larger than sample dimensions L, i.e. $\ell \gg L$, travel through the lattice without scattering and therefore can show interference effects.

In 1959, Aharonov and Bohm proposed that an electron wave in a solid has a phase factor which could be controlled by a magnetic field. This phenomenon was proved by Webb et al. in 1985 at IBM [1] in a structure similar to that shown in Figure 7.3(a) consisting of a metallic ring of diameter 800 nm made of a wire about 50 nm thick. The electrons entering the ring at P from the left have their wave function amplitude divided in two equal parts, each one travelling through a different arm of the ring. When the waves reach the exit at Q, they can interfere. Suppose that a magnetic flux Φ produced by a solenoid passes through a region inside the ring and concentric to it. We choose this highly symmetric configuration to make the calculations easier.

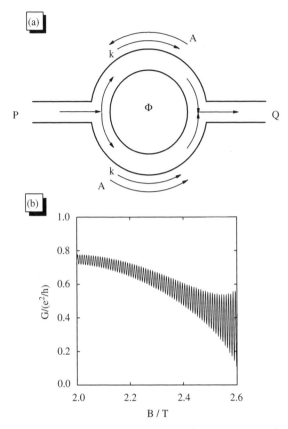

Figure 7.3. (a) Schematics of an interferometric mesoscopic system to experimentally demonstrate the Aharanov–Bohm effect; (b) conductance oscillations as a function of magnetic field due to the Aharanov–Bohm effect. After [2].

For an electron in a magnetic field $\vec{B}$, its momentum $\vec{p} = \hbar\vec{k}$ should be substituted by $\vec{p} + e\vec{A}$ where $\vec{A}$ is the vector potential ($\vec{B} = curl\,\vec{A}$). As the electron moves from P to Q in Figure 7.3(a), it is known from quantum mechanics that the change in phase is given by

$$\vartheta(\vec{r}) = \frac{e}{\hbar}\int_{P}^{Q} \vec{A}\cdot d\vec{s} \tag{7.10}$$

where the integration is the line integral along a given path which joints P with Q. The difference in phase between a wave travelling around the upper path and the lower one in

Figure 7.3(a) is:

$$\Delta\vartheta = \vartheta_1 - \vartheta_2 = \frac{e}{\hbar}\left[\int\limits_{\text{lower arm}} \vec{A}\cdot d\vec{s} - \int\limits_{\text{upper arm}} \vec{A}\cdot d\vec{s}\right] = \frac{e}{\hbar}\int\limits_{\text{circle}} \vec{A}\cdot d\vec{s} \qquad (7.11)$$

since in the top and bottom branches the electron waves advance in opposite directions to $\vec{A}$, which in our geometry is along concentric lines. Applying Stoke's theorem to Eq. (7.11) we get

$$\Delta\vartheta = \frac{e}{\hbar}\Phi \qquad (7.12)$$

where

$$\Phi = \int\limits_{\text{area}} (\text{curl}\vec{A})\cdot d\vec{S} = \int\limits_{\text{circle}} \vec{B}\cdot d\vec{S} \qquad (7.13)$$

The quantity

$$\Phi_0 = \frac{h}{e} \qquad (7.14)$$

is defined as the *quantum of flux*. We finally get for $\Delta\vartheta$ from Eq. (7.12)

$$\Delta\vartheta = \vartheta_1 - \vartheta_2 = \frac{2\pi\,\Phi}{\Phi_0} \qquad (7.15)$$

The intensity of the interference of the waves, $\psi_i \propto \exp(i\vartheta_i)$, is proportional to the probability amplitude given by

$$P = (\psi_1 + \psi_2)^2 \propto \cos(\vartheta_1 - \vartheta_2) = \cos 2\pi\frac{\Phi}{\Phi_0} \qquad (7.16)$$

and therefore *interference effects* should be observed when Φ is varied.

According to the above result, one should observe a complete oscillation when the magnetic flux Φ through the inside of the structure is changed by one magnetic quantum flux Φ_0. Since the flux area is fixed, we can appreciate from the expression for Φ, in Eq. (7.13), that when $\vec{B}$ is varied there would be oscillations in observable quantities such as the conductance. Figure 7.3(b) shows a pattern of *conductance oscillations* observed by Ford et al. (1994) [2].

One very interesting aspect of the Aharonov–Bohm effect is the observation that variations in phase can be induced by changing $\vec{B}$, even if the electron waves are not directly subjected to the action of $\vec{B}$. This point which had been controversial for some time is now settled because, contrary to the field, a vector potential $\vec{A}$ indeed exists in the region around the ring of Figure 7.3(a), and the changes in phase are produced by $\vec{A}$ according to Eq. (7.11). Another interesting observation is that the Aharonov–Bohm quantum interference effects are frequently observed even in samples of size in the micrometre range.

7.6. THE SHUBNIKOV–DE HAAS EFFECT

In this section we continue with the study of the effect of magnetic fields on the electronic and transport properties of the 2D systems. We have seen in the previous section that as the intensity of $\vec{B}$ is varied, the energy and degeneracy of the Landau levels also varies, something that should have profound effects on the transport properties of materials in general and of the 2D systems in particular. In many experimental conditions, the density in energy of electrons, n_{2D}, in the 2D system is kept constant, while the magnitude of the magnetic field is varied. As B increases, the Landau levels move up in energy since the separation between them, $\hbar\omega_c$, gets larger. Similarly, the degeneracy D of each level also increases, according to Eq. (7.9). A *filling factor* v is usually defined as

$$v = \frac{n_{2D}}{D} = \frac{2\pi\hbar n_{2D}}{eB} \tag{7.17}$$

i.e. the filling factor is equal to the quotient between the density in energy of the electrons divided by the degeneracy of each level. In general v is not an integer, but at $T = 0\,\mathrm{K}$ the maximum integer number, smaller than v, should indicate the number N of Landau levels which are completely occupied. Evidently, the top Landau level is in general partially occupied. However, when the filling factor is an integer, all Landau levels are completely occupied. Therefore, if $v = N$, then the values B_N of the magnetic field for which full occupancy of the Landau levels takes place, should be given, from Eq. (7.17), by

$$B_N = \frac{1}{N}\frac{2\pi\hbar n_{2D}}{e}, \quad (N = 1, 2, \ldots) \tag{7.18}$$

Let us now look at the variation, as a function of B, of the Fermi level, E_F, at $T = 0\,\mathrm{K}$. Qualitatively, we can say that if the filling factor is an integer, i.e. $v = N$, E_F should lie in the gap between Landau levels in which, from Figure 7.2, $n_{2D} \approx 0$. In this case, the values of the magnetic fields B_N are given by Eq. (7.18). Since $n_{2D} \approx 0$ in a fairly large energy interval, small energy changes should not practically affect n_{2D}. In addition, in this situation the electrical conductivity of the sample should be small because carriers responsible for transport lie around E_F. In contrast, when υ is an integer plus or minus 1/2, then n_{2D} shows maxima (Figure 7.2), and E_F should have values of energy located right at the peaks. In this case, a small change in energy has a large effect in n_{2D} and the conductivity of the sample should be large. Figure 7.4 shows the oscillatory dependence on the gate voltage of the potential difference U_{PP} between two probes situated along the length of the sample, like in the inset of the figure. This voltage drop is evidently proportional to the resistivity ρ_{xx} along the sample. This arrangement also allows the measurement of the resistivity ρ_{xy} across the sample, which is proportional to the Hall voltage U_H. In the case of Figure 7.4, a magnetic field B of 18 Tesla is applied perpendicularly to the 2D structure, which is kept at $T = 1.5$ K. This figure corresponds to the results of von Klitzing et al. (1980) [3] which led to the discovery of the quantum Hall effect (QHE).

Let us now focus our attention on V_{pp} or ρ_{xx} of Figure 7.4. As discussed above, the voltage oscillations of V_{pp} are a consequence of the *Shubnikov–de Haas effect*, which is really due to the formation of Landau levels by the electrons in a magnetic field. Evidently,

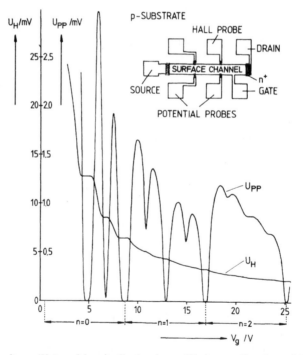

Figure 7.4. Hall voltage (V_H) and longitudinal voltage (V_{PP}), as a function of the gate voltage, for a 2D electron system in the channel of a silicon MOSFET ($T = 1.5$ K), under the action of a 18 T magnetic field (these results led to the discovery of the quantum Hall effect). After [3].

there are also oscillations in conductance corresponding to the longitudinal resistivity oscillations. In this experiment, the filling factor of the Landau levels is changed by the positive gate voltage, which controls the electron concentration at the Si–SiO$_2$ interface. It is also interesting to point out that the Shubnikov–de Haas oscillations in the 2D systems depend only on the component of $\vec{B}$ perpendicular to the interface. These oscillations were previously observed in bulk semiconductors, but they were much weaker and dependent on both components of $\vec{B}$, the perpendicular and the in-plane ones.

7.7. THE QUANTUM HALL EFFECT

7.7.1. *Experimental facts and elementary theory of the integer quantum Hall effect (IQHE)*

One striking observation in the results of Figure 7.4 comes from the quantification of the values of the Hall voltage V_H (in the figure, U_H). It is usual when commenting on these results, to speak in terms of the Hall resistance R_H, but other authors speak in terms of the transversal or Hall resistivity ρ_{xy}. Since the voltage steps have the same value, independent of the shape of the sample, we can speak indistinctively in terms of the resistivity or the resistance. The subindexes of the resistivity are related to the fact that the difference in voltage is measured in the perpendicular direction (y) to the current flow (x). Evidently the longitudinal resistivity ρ_{xx} is obtained when the voltage differences are recorded across the direction of the current.

The *quantification of the values of the Hall resistance* is given, with an outstanding precision, by the equation:

$$R_H = \frac{h}{e^2}\frac{1}{n} = 25812.807\ \Omega \left(\frac{1}{n}\right), \quad n = 1, 2, \dots \tag{7.19}$$

These values of R_H can be registered very accurately, since the experimental curves for V_H in Figure 7.4 show broad "plateaux", i.e. they remain constant over a wide range, even if the gate voltage and hence the 2D electron concentration varies (later on, we will see that these "plateaux" also appear when the electrical parameters are recorded as a function of the magnetic field). It is also interesting to observe that the "plateaux" in the values of ρ_{xy} appear precisely when ρ_{xx} becomes zero, as it can be appreciated from Figure 7.4. We would also like to remark that the Hall resistance R_H should not be confused with the Hall coefficient, for which a similar nomenclature is used. In fact R_H is known today as the *von Klitzing constant* (when $n = 1$ in Eq. (7.19)) and is written as R_K. The value of R_K can be measured at present with an accuracy of the order of 10^{-9} or better and for this reason is used as a standard in metrology (see Section 7.7.4 below).

One elementary theory that explains the values of the quantification of the Hall resistance R_H of Eq. (7.19) is based on the location in energy and the degeneracy of the Landau levels, together with a classical argument for transport of carriers. In effect, in the above definition of R_H we can substitute the Hall voltage, V_H, as in the classical Hall effect by bBv (b is the width of the sample and v the carrier drift velocity) and the current, I, by $bn_{2D}ev$. With these substitutions, the expression for the quantized resistance is:

$$R_H = \frac{V_H}{I} = \frac{B}{en_{2D}} \qquad (7.20)$$

For a given distribution, when the last Landau level becomes completely filled, the next one is completely empty, and there cannot be electron scattering. Substituting in Eq. (7.20) the values of the magnetic field B_N when the levels are completely filled, Eq. (7.18), one obtains for R_H:

$$R_H = \frac{1}{n}\frac{h}{e^2}, \quad n = 1, 2, 3, \dots \qquad (7.21)$$

Observe also that the inverse of R_H gives:

$$(R_H)^{-1} = \frac{e^2}{h}n \qquad (7.22)$$

which is in accordance with the values obtained for the quantization of conductance (Section 6.4.1).

Another interesting observation is that the quantized values of R_H correspond with the null values of the longitudinal resistance, as shown in Figure 7.4. This is interpreted as the filling of Landau levels as observed when we studied the Shubnikov–de Haas oscillations in Section 7.6. The results described in this Section constitute the so-called *integer quantum Hall effect (IQHE)*.

7.7.2.   Edge states and the IQHE

The IQHE has also been interpreted by Büttiker in the late 1980s in terms of the Landauer–Büttiker formalism [4] for multi-probe quantum transport introduced in Section 6.4.2. Let us consider that, to the various Landau levels of the 2D electron system, correspond classical cyclotron orbits caused by the perpendicular magnetic field. For simplicity, let us assume that we have a sample of longitudinal shape, with electrical leads connected to each side. The cyclotron orbits are shown in Figure 7.5(a), directed counterclockwise. Before we proceed, the reader should be cautioned that the argument which follows is mainly qualitative; however, a detailed discussion can be found in the specialized

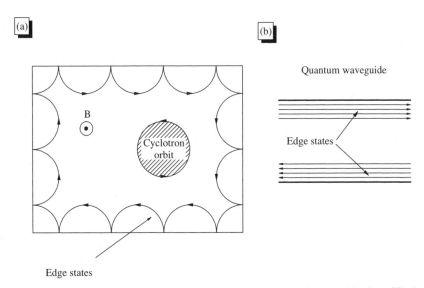

Figure 7.5. (a) Skipping orbits corresponding to edge states, and cyclotron orbits for a 2D electron system under the action of a magnetic field; (b) schematic representation of the edge currents.

literature [5]. The closed cyclotron orbits, which do not carry current on the average, are no longer possible near the edges of the sample. Therefore, as observed in Figure 7.5(a), the electrons at the edges move in the so-called *skipping orbits*, which have a net drift velocity, and consequently are the origin of the edge currents. Quantum mechanically, the states associated with the skipping orbits are called *edge states*. The upper edge states in Figure 7.5(a) have a positive velocity, while the lower ones have negative velocity. These edge states, which propagate in opposite directions, are usually represented by compact flow lines of edge currents, localized in a quantum waveguide (Figure 7.5(b)).

We have seen in Section 6.4 that quantization of conductance, and therefore resistance, was derived in a natural manner from the Landauer–Büttiker formalism of quantum transport. Similarly, the IQHE can be interpreted on the basis of edge channels. For this, we consider a sample with a bar geometry, such as the ones used to detect the IQHE and apply the general formula of Eq. (6.17) of Chapter 6 to evaluate the current transported by the edge channels. In the application of this equation, which we reproduce here for convenience:

$$I_i = \frac{2e^2}{h} \left[(N_i - R_i) V_i - \sum_{j \neq i} T_{ij} V_j \right] \qquad (7.23)$$

we should strictly follow the line of argumentation introduced by Landauer and Büttiker. Let us assume that the edge current contains N channels, although in Figure 7.6 we have represented only two. Current only flows in or out of the sample through the contact leads 1 and 4 and the Hall voltage arises between probe contacts 6 and 2 or, alternatively, 5 and 3. The longitudinal resistance of the sample (see Section 7.7.1) can be measured between contacts 5 and 6 or 3 and 2. As indicated by the edge channels of Figure 7.6, the current arising at contact 1 enters into probe 6, but since this is a voltage probe, it cannot take net current; therefore a current of the same value enters the probe, so that $I_6 = 0$. The same argument can be applied to the other voltage probes 2, 3, and 5. For the application of Eq. (7.23) for the currents, we assume perfect contacts, i.e. they do not reflect currents which means $R_i = 0$. In Figure 7.6 we also observe that N states propagate from contact 1 to contact 2, i.e. $T_{21} = N$, but no states propagate from contact 2 to contact 1, i.e. $T_{12} = 0$. Evidently the same applies to the other adjacent contacts and therefore $T_{32} = T_{43} = T_{54} = T_{65} = T_{16} = N$ and $T_{23} = T_{34} = T_{45} = T_{56} = T_{61} = 0$. All remaining T_{ij} are zero since currents cannot jump contacts as seen in Figure 7.6. With all

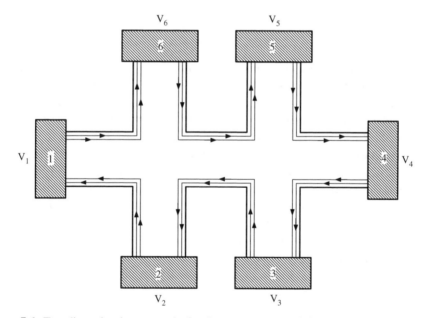

Figure 7.6. Two-dimensional test sample for the measurement of the quantum Hall effect. The current goes from probe 1 to 4. The Hall voltage can be measured from probes 6 and 2 or, alternatively, 5 and 3. The voltage drop in the direction of the current is measured from probes 5 and 6 or 3 and 2. The edge currents (two in the figure) are also shown.

these considerations, let us write Eq. (7.23) in matrix form for compactness:

$$
\begin{bmatrix} I_1 \\ I_2 \\ I_3 \\ I_4 \\ I_5 \\ I_6 \end{bmatrix} = \begin{bmatrix} -I \\ 0 \\ 0 \\ I \\ 0 \\ 0 \end{bmatrix} = \frac{Ne^2}{h} \begin{bmatrix} 1 & 0 & 0 & 0 & 0 & 0 \\ 1 & 1 & 0 & 0 & 0 & 0 \\ 0 & 1 & 1 & 0 & 0 & 0 \\ 0 & 0 & 1 & 1 & 0 & 0 \\ 0 & 0 & 0 & 1 & 1 & 0 \\ 1 & 0 & 0 & 0 & 1 & 1 \end{bmatrix} \begin{bmatrix} V_1 \\ V_2 \\ V_3 \\ V_4 \\ V_5 \\ V_6 \end{bmatrix} \tag{7.24}
$$

We have written $I_1 = -I$ and $I_4 = I$ since the arrows in Figure 7.6 indicate the electron motion. Several interesting conclusions can be reached from the above set of equations. From the equations for I_2 and I_3, it can be obtained that $V_2 = V_3 = V_4$, and similarly $V_1 = V_5 = V_6$, i.e. all the contacts along the top edge are at the source potential, while the ones at the bottom edge are at the drain potential.

In order to calculate the Hall resistance we have to divide the voltage between probes 6 and 2 by the current I between contacts 1 and 4:

$$
R_{14,62} = \frac{V_6 - V_2}{I} \tag{7.25}
$$

Substituting V_6 and V_2 by their expressions obtained from Eq. (7.24), we get for the Hall resistance:

$$
R_{\mathrm{H}} = \frac{h}{e^2} \frac{1}{N} \tag{7.26}
$$

Proceeding in the same fashion for the longitudinal resistance, we can calculate the voltage difference between probes 5 and 6, for instance, and one obtains zero values, as expected. The above results show that the quantization of the transverse resistance, and therefore conductance, is consistent with the quantization of conductance seen in Section 6.4.1.

7.7.3. *Extended and localized states*

The above models put forward for the explanation of the IQHE predict correctly the quantization of the Hall resistivity ρ_{xy}, simultaneously with the null values of the longitudinal resistivity ρ_{xx}. However, these models do not explain the existence of the Hall "plateaux", i.e. the constancy of the Hall voltage or Hall resistance in a given range of values of the magnetic field. The existence of Hall "plateaux" of finite width can be interpreted in terms of localization of the electron states as a consequence of disorder. We have already appealed to disorder while explaining the broadening of the Landau levels in Section 7.4. It is known from solid state physics that, in addition to *Bloch extended states*, there can be also *localized states* originated by disorder, as explained by the *Anderson localization theory* (1958) [6]. The degree of localization can be characterized by the localization length α, defined in Section 1.3. In fact, according to this theory, for sufficiently disordered semiconductors, all states in a 3D solid are localized. A partial degree of localization can be caused by point defects such as dopant impurities, structural disorder which causes fluctuations in the electrostatic ionic potentials and interface roughness scattering (Section 6.2.1) in the case of heterojunctions.

Figure 7.7 shows the DOS function, for which the δ-functions have been broadened as explained in Section 7.4 and Figure 7.2. Now there are extended Bloch states around the peaks at energies $E = (n + 1/2)\hbar\omega_c$ which represent the mobile electrons and localized non-conducting states. This figure also shows the mobility edges separating the extended and localized states. In this situation, as remarked by Laughlin (1981) [7], the Landau levels are not filled at fixed concentrations. When the Fermi level lies within the localized states, they do not participate in conduction, only the extended states do. In this case, theoretical calculations show that the extended states have to carry more current to compensate for the unlocalized states and to maintain constant the value of the resistivity as shown in Figure 7.4. The additional current is due to the acceleration caused by the scattering potentials related to the disorder. At this point, it might be interesting to comment that although the IQHE was discovered more than two decades ago, there are still some theoretical aspects under intense investigation.

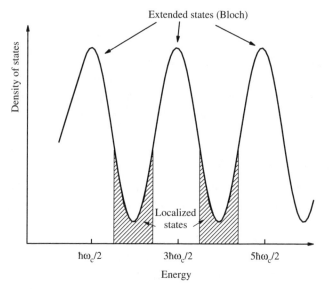

Figure 7.7. Schematic representation of the extended and localized states in the DOS function of a 2D electron system in a magnetic field.

7.7.4. Applications in metrology

The advantage of the modern standards adopted by the International System of Units (SI), based on fundamental constants, is that they do not depend on ambient conditions and can be realized at any location in the world. For instance, the unit of time (second) is now defined in terms of the period of the radiation of atomic transitions in atomic Cs.

One of the most significant applications of the QHE is in the field of metrology. In particular, since 1990, the *standard of resistance* [8] has been adopted by measuring the Hall "plateaux" in ρ_{xy}. As indicated in Section 7.7.1, this standard is the von Klitzing constant defined by Eq. (7.19) with $n = 1$:

$$R_{\text{K}} = \frac{h}{e^2} = 25812.807572 \,\Omega \tag{7.27}$$

The specific number of ohms given by Eq. (7.27) was recommended by a consultant international committee after analysing the best data available until 1998. This number has an improvement of about three orders of magnitude with respect to the first standard of 1990. The resolution of the measurements is of a few parts in 10^{10}, so the precision of the value of the resistance in Eq. (7.27) is higher than 10^{-9}, at least two orders of magnitude better than the old realization of the ohm based on standard resistors.

The impressive accuracy in the determination of R_{K} rests on several facts. First on the existence of the Hall "plateaux", which remain extremely constant in relatively wide ranges of variations of both the magnetic fields and gate voltages, as in the case of the measurements of Figure 7.4 in MOSFET–Si structures. Second, it is important to remark that the value of $R_{\text{K}}(\equiv h/e^2)$ does not depend on material properties, although in practice only two types of samples are used, one based on the Si–MOSFET and the other on the AlGaAs/GaAs heterostructure. Third, the values do not depend on geometrical factors such as the width of the test bar samples. In fact, it has been demonstrated that size effects are negligible in the test samples used today for metrology purposes. However, extreme care has to be taken with the electrical contacts, as well as with the current density which should not exceed a critical value or otherwise the IQHE breaks down. The precise conditions for the breakdown of the IQHE for high density currents are still under intense theoretical and experimental research.

The fine structure constant α, defined by

$$\alpha = \frac{e^2}{2\varepsilon_0 ch} \tag{7.28}$$

constitutes a unique combination of fundamental constants such as the electron charge, speed of light, Planck's constant, and permittivity of vacuum. From Eqs (7.27) and (7.28),

α is related to the von Klitzing constant R_K by

$$R_K = \frac{h}{e^2} = \frac{1}{2\varepsilon_0 c\alpha} \tag{7.29}$$

Therefore, the value of the fine structure constant, which is the basis of the fundamental concepts of physics, can be obtained at present by means of the IQHE. The values of α obtained by measuring R_K are comparable, if not better, than the theoretical calculations of quantum electrodynamics (QED) which have an estimated uncertainty of 1 part in 10^9. The technique based on the IQHE also yields more accurate values of α than those determined experimentally through the measurement of the anomalous magnetic moment of electrons.

7.7.5. The fractional quantum Hall effect (FQHE)

As we recall from previous sections, in order to observe the IQHE, samples do not have to be extremely pure or have a perfect crystallinity. In fact, we had to invoke some degree of disorder, when we introduced the concept of "localized states" (see Section 7.7.3) to explain the observed Hall voltage "plateaux". Two years after the discovery of the IQHE, Tsui, Stormer, and Gossard (1982) discovered the so-called *fractional quantum Hall effect (FQHE)* [9]. As is observed in Figure 7.8, in the FQHE new "plateaux" associated with ρ_{xy} as well as the null values for ρ_{xx} appear at fractional values of the filling factor ν given by Eq. (7.17). These fractional values are of the form $\nu = p/q$ where p and q are integers and q is odd. The FQHE is observed in the 2D systems formed at the heterojunctions of very pure semiconductor samples, and consequently with high electron mobility, and low 2D electron concentration. In addition, the temperature should be close to $0\,K$ and the magnetic field should be high enough, as to allow only the existence of one Landau level. The filling factor is then equal or smaller than one; therefore all electrons are at the same energy which is the lowest and is given, according to Eq. (7.8), by $\hbar\omega_c/2$. As can be appreciated in Figure 7.8, strong features appear for the filling factors $\nu = 1/3, 2/3, 2/5, 3/5, \ldots$

It can be observed from the above experimental facts that the FQHE is a completely different phenomenon than the IQHE. Therefore, if the latter is well explained by a model based on the non-interacting electrons, it is reasonable to expect that many-body interactions among the electrons should now be taken into account. Accordingly, soon after the discovery of the FQHE, it was suggested that the electron gas could undergo a condensation into a *Wigner crystal*, in which the energy related to the Coulomb repulsions is minimized. However, this idea was soon abandoned, after theoretical calculations showed that values of the total energy lower than that of a Wigner crystal could be obtained for electronic systems of a few electrons, under the action of very strong magnetic fields.

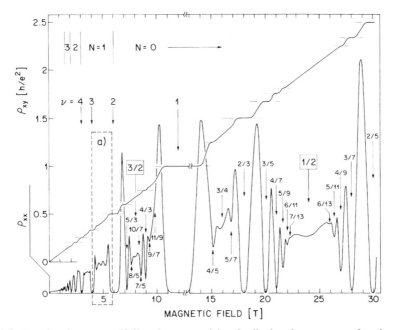

Figure 7.8. Fractional quantum Hall resistance and longitudinal resistance, as a function of the magnetic field, for a 2D electron system in a modulation-doped heterojunction AlGaAs–GaAs ($T = 0.085$ K). These results led to the discovery of the fractional quantum Hall effect. After [9].

In fact, we have already seen that for very high values of B, all Landau levels of a 2D electron system collapse to one ground level.

Soon after the discovery of the FQHE, Laughlin (1983) [10] proposed a theory based on many-body Coulomb interactions between the electrons at very high magnetic fields and low temperatures. According to this theory, the electron gas condenses into a correlated quantum fluid, where electrons are kept apart from each other due to the Coulomb interaction. Laughlin calculated the ground state wave function of the many-body electron system and found that this state was separated by an energy gap from the excited states. In this theory, Laughlin found that the correlated ground state was formed when the Landau level is partially filled according to a fractional value of v, for example, 1/3. In addition, he showed that the elementary excitations of the system correspond to particles of fractional charge given by 1/3 of the electron charge. Recent experiments, based on the measurement of electrical noise in the current passing through quantum point contacts, seem to confirm the existence of *fractional electronic charges*.

One of the most attractive theories to explain the FQHE is based on the model of the *composite fermion (CF)*, introduced by Jain (1989) [11] and later expanded by

Halperin et al. (1993) [12]. According to this model, two magnetic flux tubes of quanta Φ_0 are attached to each electron. One can think of these flux tubes as infinitely small solenoids carrying the quanta of flux Φ_0 given by Eq. (7.14). This unit behaves as a fermion, hence the name of "composite fermion". Looking carefully at Figure 7.8 it can be observed that around the value of the filling factor $\nu = 1/2$, the longitudinal resistivity ρ_{xx} seems to behave normally, as if B was not present. It can be shown that for $\nu = 1/2$ and its corresponding value of $B = B_0$, the CF behaves as if the magnetic field did not exist, while for values of B larger or smaller than B_0, then the CF obeys to the existing magnetic fields. Based on the model of the CF, Jain was able to explain most of the series of fractional values of $\nu = p/q$ for which the FQHE is manifested.

REFERENCES

[1] Webb, R.A., Washburn, S., Umbach, C.P. & Laibowitz, R.G. (1965) *Phys. Rev. Lett.*, **54**, 2696.

[2] Ford, C.J.B., Simpson, P.J., Zeiber, I., Franklin, J.D.F., Barnes, C.H.W., Frost, J.E.F., Ritchie, D.A. & Pepper. M. (1994) *Journal of Physics: Condensed Matter*, **6**, L725.

[3] von Klitzing, K., Dorda, G. & Pepper, M. (1980) *Phys. Rev. Lett.*, **45**, 494.

[4] Büttiker, M. (1988) *Phys. Rev. B*, **38**, 9375.

[5] *The Quantum Hall Effect* (1990) Eds. Prange, R.E. and Girvin S.M. (Springer-Verlag, New York).

[6] Anderson, P.W. (1958) *Phys. Rev.*, **109**, 1492.

[7] Laughlin, R.B. (1981) *Phys. Rev. B*, **23**, 5632.

[8] Jeckelmann, B. & Jeanneret, B. (2001) *Rep. Prog. Phys.*, **64**, 1603.

[9] Willet, R., Eisenstein, J.P., Stormer, H.L., Tsui, D.C., Hwang, J.C.M. & Gossard, A.C. (1987) *Phys. Rev. Lett.*, **59**, 1776.

[10] Laughlin, R.B. (1983) *Phys. Rev. Lett.*, **50**, 1395.

[11] Jain, J.K. (1990) *Phys. Rev. B*, **41**, 7653.

[12] Halperin, B.I., Lee, P.A. & Read, N. (1993) *Phys. Rev. B*, **47**, 7312.

[13] von Klitzing, K. (1984) *Physica*, **126**, B-C, 242.

FURTHER READING

Ferry, D.K. & Goodnick, S.M. (1997) *Transport in Nanostructures* (Cambridge University Press, Cambridge).

Hamaguchi, Ch. (2001) *Basic Semiconductor Physics* (Springer, Berlin).

Imry, Y. (1977) *Introduction to Mesoscopic Physics* (Oxford University Press, Oxford).

Murayama, Y. (2001) *Mesoscopic Systems* (Wiley-VCH, Weinheim).

Schäfer, W. & Wegener, M. (2002) *Semiconductor Optics and Transport Phenomena* (Springer, Berlin).

PROBLEMS

1. **Similarity of the Aharonov–Bohm effect and interference effects in SQUIDS.**
   (a) Discuss the similarity between interference effects in Superconducting Quantum Interference Devices (SQUIDS) and the Aharonov–Bohm effect (Section 7.5). *Hint*: recall that the SQUID critical current is modulated by the quantum flux Φ_0, yielding a Fraunhoffer interference pattern. (b) Given the value of Φ_0 and the fact that SQUIDS can have areas of about $1\,\text{cm}^2$, compare the accuracy in the determination of changes in B (measured in Tesla) by counting cycles in current with the accuracy of measuring resistances discussed in Section 7.7.4. (c) Discuss similarities between phase coherence of a superconductor macroscopic wave function and the coherence of phases introduced in Sections 7.5 and 1.4.

2. **Shubnikov–de Haas oscillations.** (a) Show that for the Shubnikov–de Haas oscillations observed in a 2D electron systems the period in $1/B$ for the oscillations is given by $\Delta(1/B) = 2e/hN_s$, where N_s is the electron concentration. (b) From the results on the integer quantum Hall effect of Figure 7.4, calculate approximately, the 2D carrier density.

3. **Integer quantum Hall effect (IQHE).** (a) From the values for the ρ_T "plateaux" of Figure 7.4, show that they are in agreement with the Eq. (7.19), i.e.:

$$R_H = \frac{h}{e^2}\frac{1}{n} = 25812.807\,\Omega\left(\frac{1}{n}\right), \quad n = 1, 2, \ldots$$

   (b) From the variation of ρ_L as a function of B from Ref. [13] find the 2D concentration of electrons (*Hint*: use low values of B since for high values, more than one subband might be populated).

4. **Variation of the Fermi level with B and the IQHE.** A magnetic field B is applied perpendicularly to a planar heterojunction similar to those employed to observe the integer quantum Hall effect (IQHE). Suppose that the n_{2D} electron concentration is $n_{2D} = 10^{12}\,\text{cm}^{-2}$ and the temperature very low so that $kT \ll \hbar w_c$, where $w_c = eB/m_e^*$ is the cyclotron frequency. (a) Find the dependence of the Fermi energy E_F as a function of B (call E_F^0 the Fermi energy when $B = 0$). *Hint*: at very low temperatures, if n Landau levels are exactly filled, E_F lies in the empty space of Figure 7.2 between the energies of the levels n and $n + 1$, or in the level $n + 1$ if this level is partly occupied. As B increases, E_F also increases linearly with B until

it reaches a value so that the next level becomes empty (the degeneracy of the levels and the fill factor increase with B) and E_F falls back to the previous level. As a result, E_F oscillates around E_F^0. (Note that this behaviour of E_F is what really makes ρ_{xx} oscillate.) (b) Estimate the maximum temperature at with the steps in ρ_T of Figure 7.4 would not be observed.

5. **2D electron system under high magnetic fields**. Using the expressions of problem 1.5 for the relations between the 2D resistivity and conductivity tensors, as well as the expressions for the velocities and R_H derived in problem 1.6, show that, under high magnetic fields as those necessary for the observation of the IQHE, we have: $\rho_{xx} \approx \sigma_{xy}/\sigma_{xy}^2$ and $\rho_{xy} \approx 1/\sigma_{xy} = R_H B$. Observe that according to this result ρ_{xx} is proportional to σ_{xx}. Please, comment on the origin of this unexpected result.

6. **Spin splitting of Landau levels in a 2D electron system**. (a) Show that to each of the energies found for the Landau levels in Section 7.2, one should add the spin splitting energy. Therefore, the Landau levels have in reality, the energies:

$$E_n = \frac{\hbar e B}{m_e^*}\left(n + \frac{1}{2}\right) \pm \frac{1}{2}g^*\mu_B B, \quad n = 0, 1, 2, \ldots$$

where g^* is the effective g factor and μ_B is the Bohr magneton. (Observe that although we have twice the number of Landau levels, if we account for the spin splitting, the derivations of Sections 7.6 and 7.7 do not change because, simultaneously the degeneracy is half the previous value.) (b) Calculate the value of $g^*\mu_B B$ and compare it with the difference in energy between Landau levels.

7. **Effect of subband occupation in the IQHE**. Indicate in a schematic plot of ρ_L vs B, the effect of the occupation of two subbands $n = 1$ and $n = 2$. (*Hint*: use the expression $\Delta(1/B) = 2e/hn_{2D}$ for the two values of n_{2D}.) Discuss what is the situation of the oscillations of ρ_L vs B both for low and high values of B.

Chapter 8

Optical and Electro-optical Processes in Quantum Heterostructures

Chapter 8

Optical and Electro-optical Processes in Quantum Heterostructures

8.1. INTRODUCTION

The determination of the optical properties of low-dimensional semiconductor structures allows the verification of the physics we have developed in Chapters 1, 4, and 5 on quantum wells (QWs), superlattices, and quantum dots (QDs). The agreement between the theory and the experimental results shows that the predictions derived from the energy dependence of the DOS function, and the behaviour of excitons in 2D and 0D systems, were indeed correct. The optical properties of low-dimensional quantum heterostructures that we study in this chapter constitute the basis of a new generation of electronic devices, such as QW lasers, QD lasers, IR photodetectors, electro-optic modulators, etc. The operation of these devices, many of them in commercial production at present, will be treated in Chapter 10.

In Section 8.2, we review interband and intraband optical transitions in quantum wells and the influence of the excitonic effects. In this section, the studies on optical properties of single quantum well are extended also to superlattices. Section 8.3 is dedicated to the optical properties of QDs and nanocrystals, in which the confinement on the three dimensions in 0D structures leads to new phenomena with applications in newly developed QD lasers and photodetectors. The section starts with a review of the growth of QD and nanocrystals, since their optical properties are strongly dependent on the growth technique.

Section 8.4 deals with the effects of electric fields in quantum wells. The effects of an electric field on the optical properties (index of refraction, transmission spectra, etc.) are very much enlarged in low-dimensional systems. Modern electro-optical modulators, based on the quantum-confined Stark effect (QCSE), take advantage of the large changes in excitonic optical absorption as the electric field is varied. The effects of electric fields on superlattices are treated in Section 8.5. In superlattices, the electric field produces localization of carriers at the wells and generates a set of equidistant energy levels (Stark ladders). The recent (1993) observation of Bloch oscillations and the related microwave emission are also treated in Section 8.5.

8.2. OPTICAL PROPERTIES OF QUANTUM WELLS AND SUPERLATTICES

The optical properties related to interband transitions in quantum wells are quite different than those corresponding to bulk semiconductors, since one has to consider both the 2D optical density of states and the fact that excitonic absorption is much stronger in 2D systems (Chapter 4). With respect to intraband transitions, in contrast to the 3D situation, 2D systems can show transitions without the necessity of involving phonons. Intraband transitions can be among electrons (or holes) in confined states in wells, or between confined states and the continuum. These transitions can be tailored for light emission (quantum cascade lasers) or detection (IR photodetectors).

In Figure 8.1, we have represented a quantum well both in real and wave vector spaces. *Interband transitions* take place from an initial state in the valence band to a final state in the conduction band. First, observe that absorption will appear at energies higher than for the 3D case, since the energy difference between these states is larger than the energy bandgap of the semiconductor. In order to calculate the transition rates, we should follow the procedure of the Fermi Golden rule, Eq. (2.26) of Chapter 2, for time-dependent perturbation theory:

$$W = \frac{2\pi}{\hbar}\rho(E)\left|H'_{nk}\right|^2 \tag{8.1}$$

with the perturbation, associated with the interaction of photon–electron, given by:

$$H' = -e\vec{r} \cdot \vec{E}_0 \tag{8.2}$$

where $\vec{r}$ is the position vector of the electron in the plane of the interface, i.e. $\vec{r} = (r_x, r_y)$ and $\vec{E}_0$ is the amplitude of the electric field related to the incident light. The matrix element

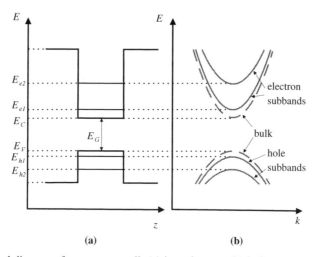

Figure 8.1. Band diagram of a quantum well: (a) in real space; (b) in k-space.

for the transitions between the initial electron state Ψ_i in the valence band and the final state Ψ_f in the conduction band is:

$$M = \int \Psi_f^* \, r_x \, \Psi_i \, dr \tag{8.3}$$

Observe that since $\vec{r} = (r_x, r_y)$ is a planar vector, we can write either r_x or r_y in Eq. (8.2) which are in reality the (x, y) coordinates of the electron in the plane. Assuming cubic (square) symmetry, the directions x and y are equivalent. In the crystalline 2D structure the electron states are described by Bloch functions:

$$\Psi_i \propto u_v\,(\vec{r})\,\psi_{vn}\,(z)\,e^{i\vec{k}\cdot\vec{r}} \tag{8.4}$$

$$\Psi_f \propto u_c\,(\vec{r})\,\psi_{cn'}\,(z)\,e^{i\vec{k}\cdot\vec{r}} \tag{8.5}$$

where the vectors $\vec{r}$ and $\vec{k}$ correspond to the electrons in the 2D system. We have written the same wave vector in Eqs (8.4) and (8.5) because there is momentum conservation in the transition and the momentum of the photons is negligible in comparison to that of the electrons. By the substitution of Eqs (8.4) and (8.5) in Eq (8.3), the matrix element M can be expressed as the product of two factors, i.e.:

$$M = M_{vc} M_{nn'} \tag{8.6}$$

where:

$$M_{vc} = \int u_c^*(\vec{r}) r_x u_v(\vec{r}) d\vec{r} \tag{8.7}$$

is the matrix element corresponding to the dipole moment transitions among Bloch states in the valence and conduction bands, and

$$M_{nn'} = \int_{\text{well}} \psi_{cn'}^*(z)\psi_{vn}(z)dz \tag{8.8}$$

corresponds to the overlapping between the electron and hole wave functions in the wells.

Since, according to Eq. (8.6), the matrix elements of the interband transitions decompose into two factors, the following conditions or *selection rules*, must be satisfied:

(i) The overlap of the electron and hole envelope functions given by Eq. (8.8) should be different from zero. Therefore, in the case of square wells of infinite walls, we have from the orthonormality of the wave functions:

$$\Delta n = n' - n = \delta_{nn'} \tag{8.9}$$

where n and n' are the quantum numbers corresponding to the electron and hole quantum wells, respectively.

(ii) The matrix elements between Bloch functions corresponding to the electric dipole transitions given by Eq. (8.7) should be different from zero. Since this is the case for AlGaAs/GaAs quantum wells, the expression of Eq. (8.9), related to envelope functions, can be taken, in practice, as the selection rule.

Figure 8.2 [1] shows the absorption spectrum of MQW of GaAs/AlAs at 6 K. Observe the increasing absorption with photon energy which in general follows the steps of the DOS function for 2D systems (Section 4.2). Added to this, the excitonic peaks can be clearly seen at the beginning of each step. In 2D systems, the binding energy of excitons, as well as their absorption, are very much enhanced by the confinement effects, and therefore, make excitons much easier to detect than in bulk semiconductors. At every peak in Figure 8.2, there appears the spin–orbit interaction doublet, which corresponds to the heavy and light hole valence bands characteristic of III-V compounds (Section 4.9). The transitions for $n = 1, 2, 3$ from the heavy holes (HH) and light holes (LH) to the electron states can be clearly distinguished in Figure 8.2.

With respect to *intraband transitions* the emission or absorption of photons can occur, as shown in Figure 8.3(a), between the "free" carriers within the conduction (or valence)

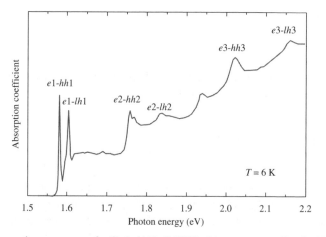

Figure 8.2. Absorption spectrum of a GaAs/AlAs MQW with quantum wells of width 7.6 nm. The transitions to the electron states can be originated at the heavy hole (HH) or at the light hole (LH) states. After [1].

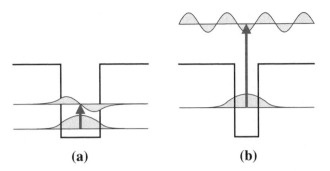

(a)   **(b)**

Figure 8.3. Intraband optical transitions in a quantum well. (a) Intersubband transitions; (b) transitions between bound states in the well and extended states. After [2].

bands, which occupy the subbands corresponding to the levels of each electron (or hole) located in the corresponding well. Notice that these *intersubband transitions* only occur within each quantum well. In addition, intraband transitions can also occur *between the quantum well states and the extended electron states* as shown in Figure 8.3(b). With a reasoning similar to the one performed in relation to interband transitions, the *selection rule* should connect now the states of opposite parity, since the matrix element now includes the position coordinate, i.e.:

$$\Delta n_z = n_{z,f} - n_{z,i} = \pm 1 \tag{8.10}$$

In an ideal case, the selection rule says Δn_z should be equal to an odd number, but when $\Delta n_z = \pm 3$ the transitions are too weak, and therefore, are not accounted for in Eq. (8.10). The other requirement is that $\vec{E}$ is perpendicular to the well, i.e. the polarization of light should occur along the z-direction. For this, in practice, the light should not be incident vertically, or alternatively a diffraction grating should be placed on the surface of the sample. Both solutions imply a partial attenuation of light and therefore to overcome it, one should use MQWs with a high number of wells (close to one hundred).

The changes induced in the optical properties of quantum wells and superlattices by electric fields are quite different, and they will be treated in Sections 8.4 and 8.5, respectively. At present, research on electro-optic effects in superlattices is centred around Bloch oscillations and the possibility of Terahertz emission (Section 8.5). However, the optical properties of *superlattices* are somewhat similar to those of quantum wells. Optical properties of superlattices, contrary to MQWs, have found few applications in optics, and the interest on them has been frequently focused on other materials science topics, such as growth of quantum dots, bandgap engineering, etc.

There are some differences between the optical properties of superlattices and quantum wells. In superlattices, the absorption edge is blue-shifted due to the splitting of the levels

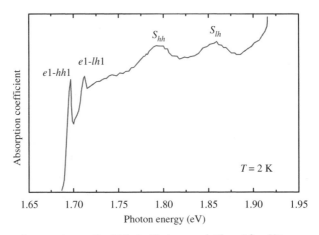

Figure 8.4. Absorption spectrum of a AlGaAs/GaAs superlattice. After [2].

into a miniband. Also the transitions are smoother, reflecting the density of states function of a superlattice (Figure 5.15) in comparison with the step function character of quantum wells. Due to the tunnelling of the electrons through the barriers, the wave function of the exciton extends along several wells in strongly coupled superlattices. Therefore, the exciton is less localized and its size and binding energies are smaller than for quantum wells. Weak excitonic features appear also near the transitions from the bottom of the hole miniband to the top of the electron miniband (saddle-point excitons). Figure 8.4 shows the absorption spectrum of an AlGaAs/GaAs superlattice [2]. Observe the similarity with the spectrum for a single quantum well. The excitons shown in the figure correspond to $n = 1$ transitions from the heavy hole and light hole to the first electron state. The peaks marked with S_{hh} and S_{lh} correspond to saddle-point excitons.

8.3. OPTICAL PROPERTIES OF QUANTUM DOTS AND NANOCRYSTALS

8.3.1. *Growth techniques. Self-assembled quantum dots*

The growth techniques of QDs are very important since they influence their structure, shape and distribution of nanocrystal sizes, stoichiometry, structure of the surfaces or interfaces, etc. Therefore, before studying in the next section the optical properties of QDs, we want first to briefly look to the growth techniques, paying special attention to the recently developed techniques of self-assembling.

Nanostructuring by *physical techniques* based on lithography and etching does not usually produce dots of sizes small enough for the detection of quantum size effects, even if one uses electron beam techniques. The advantage of these techniques is their compatibility with microelectronics techniques. One technique where advances are made very fast is scanning tunnelling microscopy (STM), and the related atomic force microscopy (AFM). Therefore, it would not be surprising if in the near future they will be used to fabricate isolated QDs as well as QD arrays.

The *chemical techniques* are the ones most used for the preparation of QDs. However, there are so many chemical techniques for producing QDs, that we will only mention a few of them:

(i) Nanocrystals in glass matrices

The techniques for growing nanocrystals inside a glass matrix are very developed, partly as a consequence of the industrial production of colour filters and photochromic glasses based on copper halides, such as CuCl, CuBr, CuI, etc. Nanocrystals of II-VI compounds (CdS, CdSe, ZnSe, etc.), starting from supersaturated viscous solutions, are also frequently incorporated in glass matrices for applications in optical filters due to the ease in controlling the dot size.

(ii) Colloidal synthesis

Colloidal synthesis, by the reduction of metal salts in solutions with organic ligands, are frequently used to produce metal nanoclusters (e.g. gold). As for II-VI semiconductors, nanoparticles of CdSe, CdS, etc. are produced from reagents containing the nanocrystals constituents. One reagent contains the metal ions (e.g. Cd^{2+}) and the other provides the chalcogenide (e.g. Se^{2-}). The size of the nanocrystals is controlled by the temperature of the solutions and the concentrations of the reagents and the stabilizers.

(iii) Gas-phase

The gas-phase methods consist of the gas condensation of clusters which precipitate on a substrate, or that are adsorbed on the substrate. The gas-phase techniques include sputtering and laser vapourization among others and are very much used for metal clusters. However, since the discovery of nanoporous silicon of visible photoluminescence, silicon clusters deposited by laser-induced decomposition of SiH_4 or magnetron sputtering, are very much investigated at present for light emission sources.

(iv) Self-assembled quantum dots

Since the 1990s, there has been great interest in *self-assembled QDs* which can be fabricated by techniques compatible with present microelectronic and optoelectronic device technologies. One fabrication technique is based on a modification of the Stranski–Krastanow (S–K) method. When a material is grown over a substrate with similar lattice constant, then the growth is monolayer by monolayer. However, if there is a lattice

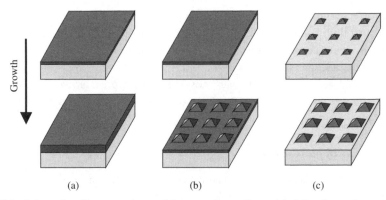

Figure 8.5. Schematic diagram of possible growth modes: (a) Monolayer by monolayer;
(b) Stranski–Krastanow (S–K); (c) Volmer–Weber (V–W).

mismatch of a small percentage, then the overlayer starts to grow monolayer by monolayer, but at a given stage the layer grows in the island mode (Stranski–Krastanow mode), so that the strain energy is minimized. For a higher lattice mismatch, the islands nucleate directly since the beginning (Volmer–Weber mode).

Figure 8.5 shows a schematic diagram of the three modes that we have mentioned: (a) the monolayer-by-monolayer growth, (b) the Stranski–Krastanow mode, and (c) the Volmer–Weber mode. The resulting mode depends not only on the lattice mismatch, but also on the values of the interface energies. QDs for lasers (Section 10.6) are grown in the S–K mode with materials like InGaAs/GaAs, but the method also works for the SiGe/Si and CdSe/ZnSe. The advantage of the S–K method over physical lithographic techniques is that the resulting QDs are smaller.

8.3.2. Optical properties

In quantum dots or boxes, the potential confines the electrons in the three spatial dimensions, which are supposed to be in the nanometre range, so that the energy levels appear as discrete bound states (Section 4.6), as in the case of isolated atoms. The confinement of the electronic wave functions has very important consequences for the optical properties.

Let us comment some general *optical properties of 0D confined systems*:

(a) Bandgap widening

The first characteristic of the optical properties of QDs is related to bandgap widening, as can be appreciated in Figure 8.1(a). Depending on the size R of the QD, which is supposed to be spherical, several regimes can be considered. To define these regimes,

R is compared with the size of the excitons as defined by its Bohr radius:

$$a_\text{B} = \frac{\hbar^2}{4\pi \varepsilon_0 \varepsilon} \frac{1}{\mu e^2} \tag{8.11}$$

where μ is the exciton reduced mass (Section 3.7.3). In the "strong confinement regime", for $R \leq 2a_\text{B}$, the confinement energy is greater than that corresponding to the Coulombic interaction. In this case, the excitonic effects due to the electron–hole interaction can be neglected, and we can consider that electrons and holes reside in quantum boxes defined by the dot dimensions. This is a consequence of the fact that the Coulombic energy increases as the inverse of R while the confinement energy increases as the square of the inverse. The excitonic Bohr radius in CdS is 29 Å, so nanocrystals of sizes smaller than about 50 Å should behave as QDs. When R is much smaller than a_B, then the effective mass approximation does not apply and the dot should be treated as a large molecule and consequently should be treated in terms of molecular orbitals.

In the "weak confinement regime", $R \geq 4a_\text{B}$, the envelope function is practically not affected, although the exciton increases its kinetic energy associated to the centre of mass motion, and therefore, there is a decrease in its binding energy. This case occurs in CdSe for R larger than about 100 Å. For copper halides (CuCl, CuBr), a_B is so small ($\sim$1 nm) that QDs are usually in the weak confinement regime. Evidently, the most difficult case to consider from a theoretical point of view is medium or intermediate confinement among the two above cases, since approximations cannot be applied.

(b) Enhancement of oscillator strengths

We have seen that as the dimensionality decreases (3D → 2D → 1D → 0D) the allowed electron states become more concentrated in energy, as shown by the respective energy dependences of the DOS functions. For the 0D limit, the dots behave like atoms with sharp energy levels and the oscillator strengths of optical transitions are larger. This is important for optoelectronic devices and we will see more clearly in Sections 10.3 and 10.6, how concentration of states on energy leads to higher gains in lasers. Due to the same reason, the electro-optic effects (Sections 8.4 and 8.5), used in quantum well optoelectronic modulators (Section 10.8), become stronger as the dimensionality decreases.

(c) Optical transitions

As we have seen in Section 8.2, intersubband optical transitions in 2D systems are only allowed when light propagates in the quantum well plane, so that the photon electric field is perpendicular to the interfaces. However, QDs can absorb incident light polarized in any direction. This is because of the confinement in the three optical directions which means that the wave functions of electrons are also quantized in the three spatial directions.

(d) Broadening of spectra

Another important property of QDs is that there is no temperature dependence of the line width of the spectra, since there are no continuum states for the electrons to be promoted. However, in 1D and 2D there are one and two directions of continuum k states, respectively. The theoretical expected line spectra of QDs are not completely sharp (δ-functions) but in reality they have some width ($\sim$10 meV), as a consequence primarily of QD size distribution. If the dots are very homogeneous, then the width can be smaller than 10 meV, but in contrast, for the cases of wide size distributions, the width can approach 0.1 eV. Variations of the QD bandgap due to small variations in composition also produce broadening. Other causes are impurities, surface or interface states, etc. In order to study the intrinsic properties of QDs and avoid the broadening effect, *single dot spectroscopy* has been developed (see below).

After describing the main characteristics related to the optical properties of QDs, let us now review some of the results on *optical characterization of various QD systems*:

(i) Compound semiconductor QDs

We will only give a few examples of QDs spectra of semiconductor compounds, since there is a wealth of results. Some of the most studied QDs, have been the II-VI nanocrystals, because of their facility of preparation and important technological applications. In many cases, II-VI compounds are introduced as dopants during the glass fusion process for the production of doped glasses. The resulting nanocrystals can be controlled in size in order to modify the glass colour. Figure 8.6 shows the absorption spectra of CdS

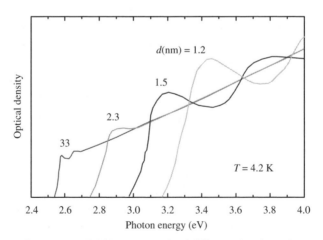

Figure 8.6. Absorption spectra of CdS nanocrystals of different sizes in a glass matrix. After [3].

($E_G = 2.6\,\text{eV}$) QDs in glass [3]. Observe the blue-shift caused by the quantum size effects. Using mixtures of Cds with another II-VI material like CdSe ($E_G = 1.75\,\text{eV}$) the whole optical spectrum can be covered. Interesting results with CdSe nanocrystals were also obtained by Empedocles et al. (1996) [4] using single-dot spectroscopy, similar to that used for single-molecule spectroscopy. Figure 8.7 shows the great difference between a single-dot spectrum of an 8 nm nanocrystal and that from an ensemble of nanocrystals.

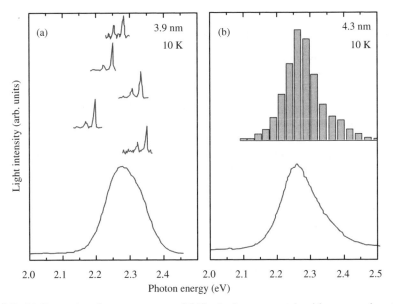

Figure 8.7. (a) Comparison between spectra of CdSe single nanocrystals with average size of 3.9 nm (top) and that of a large ensemble (bottom); (b) the histogram of the energies of about 500 dots with an average size of 4.3 nm (top) and their emission spectrum (bottom).

(ii) Self-assembled QDs

3D island formation (Section 8.3.1) was already observed in 1985 by Goldstein et al. during the growth of InAs/GaAs superlattices [5]. When a few monolayers are grown ($\sim$2.5 ML), then the InAs QD formation is produced as can be shown by high-resolution transmission electron microscopy. The photoluminescence line spectrum is quite broad due to the distribution of sizes and the photoluminescence peak goes to lower energies than the one corresponding to the superlattices, because of the larger size of the dots in comparison to the superlattice period. Since 1994, there has been a great interest in these dots for the fabrication of QD lasers from the InGaAs/GaAs system. Figure 8.8 shows the photoluminescence and electroluminescence spectra of InAs/GaAs QDs grown by MOCVD, similar to those used in QD lasers [6].

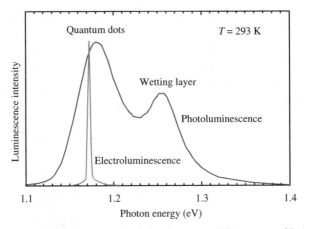

Figure 8.8. Photoluminescence (PL) and electroluminescence (EL) spectra of InAs/GaAs quantum
dots grown by MOCVD. After [6].

Grundmann et al. (1995) [7] used spatially resolved cathodoluminescence for obser-
vations of single InAs/GaAs QDs (Figure 8.9). They observed several important facts:
(a) the spectral lines do not increase their width in the temperature range up to 70 K,
which proves the above mentioned 0D character of the nanocrystals; (b) the width of the
individual lines is smaller than 0.15 meV; (c) the spectroscopic resolution of the technique

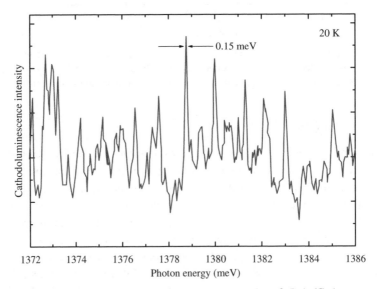

Figure 8.9. Spatially resolved cathodoluminescence spectrum of InAs/GaAs quantum dots.
After [7].

is so good that taking into account the relatively small number of molecules in each dot (about 2000), the addition or substraction of a single InAs molecule can be detected.

(iii) Indirect gap semiconductor nanocrystals

Since the discovery of a strong luminescence in Si nanocrystals by Canham (1990) [8], there has been a high interest in the research of both Si crystalline clusters and porous silicon. Although we know from semiconductor physics that indirect gap bulk semi-conductors should present a very low efficiency for the emission of light, the situation changes in a nanocrystal. The emission of light by *silicon nanocrystals* occurs in the visible ($\sim 2\,eV$) instead of the IR ($E_G = 1.1\,eV$) because of bandgap widening as a consequence of quantum size effects (Figure 8.1(a)). Numerous experiments have proved, in addition, the blue-shift effect as the size of the crystal diminishes. As for the strong luminescence, it has been usually explained in terms of the relaxation of the $\vec{k}$ wave vector (or momentum $\vec{p} = h\vec{k}$) conservation law, under low-dimensional confinement effects (see below).

Strong visible luminescence is specially observed in *porous silicon (PS)* prepared from silicon single crystals by anodic electrolytic formation in HF–ethanol electrolytes. The resulting porous layer is formed by 0D and 1D nanocrystals of nanometre range size. The emission spectrum is quite wide as a consequence of the quasi-Gaussian distribution in nanocrystal size. At the beginning many authors attributed the visible strong luminescence to the formation of siloxene ($Si_6O_3H_6$), Si-hydrides, Si-oxides, or Si-oxihydrides, incorporated on the nanocrystal surfaces and interfaces as a consequence of the electrolytic formation. However, although in some situations that could be the case, today it is known that quantum size effects are always a valid explanation, since the luminescence is still observed in nanocrystals without the above surface compounds, as checked by multitude of surface and thin film analytical techniques (Auger and XPS spectroscopies, Rutherford back-scattering spectrometry, FTIR, etc.).

Quantum size effects in Si compounds can also be observed in SiGe superlattices and quantum wells. The interest in SiGe heterostructures comes from their use in the fabrication of heterostructure bipolar transistors (Section 9.3). Strained SiGe heterojunctions are built taking advantage of the difference of bandgaps in silicon (1.1 eV) and germanium (0.74 eV), and the SiGe mixtures with values of the bandgap between the above values.

The strong luminescence observed in silicon nanocrystals and silicon–germanium superlattices can be explained by the Heisenberg uncertainty principle or by the Brillouin zone folding concept, respectively. Suppose that the Si nanocrystal has a size of about 30 nm only, or smaller. Then, the carriers, as shown in Figure 8.10(a), will have a distribution in momenta, which results in a higher probability of direct transitions. As for the case of SiGe superlattices, minibands and minigaps are formed around $k = 0$ on the diagram for the energy bands, as shown in Figure 8.10(b) and Section 5.5. As a consequence,

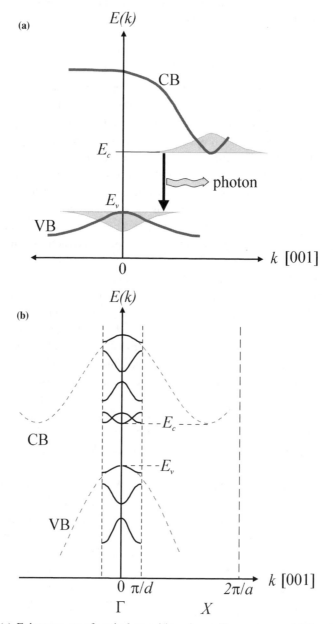

Figure 8.10. (a) Enhancement of optical transitions in a silicon nanocrystal due to Heisenberg uncertainty relation; (b) schematic illustration of the effect of Brillouin zone folding in SiGe superlattices.

the superlattice would behave as a quasi-direct gap semiconductor, with a high probability for direct optical transitions.

8.4. ELECTRO-OPTICAL EFFECTS IN QUANTUM WELLS. QUANTUM CONFINED STARK EFFECT

An external electric field induces changes in the optical properties of a material. These changes are called *electro-optical effects* and can affect the refractive index (electrorefraction effects) or the absorption coefficient (electroabsorption effects). For photon energies well below the bandgap of the material, electrorefraction is the most important effect, because the absorption coefficient is negligible. Electro-optical effects are used in many optoelectronic devices to control light with an external electronic circuit. They provide a bridge between the photonic and microelectronic devices. Semiconductor quantum wells, and all semiconductor nanostructures in general, have interesting electro-optic properties due to quantum confinement of electrons, strong exciton binding energies, and the possibility to tailor the bandgap. An electric field normal to the well plane shifts the absorption edge to lower energies (red-shift) and increases the refractive index below the absorption edge. In the proximity of the bandgap the changes are very strong and depend critically on the photon energy.

The most direct way to control the intensity of a light beam is through electroabsorption. As we know from solid state physics, an electric field applied to a bulk semiconductor shifts the absorption edge to lower energies and produces oscillations above the bandgap due to the Franz–Keldysh effect. For a light beam with a photon energy slightly below the absorption edge, the semiconductor is transparent under no applied field and absorbent under an applied field. However, in a bulk semiconductor this effect is too weak for practical applications. In quantum wells the quantum confinement produces new electroabsorption effects which are much stronger than in bulk materials (see also Section 10.8). This opens the way for the fabrication of electroabsorption modulators based on quantum wells. When the electric field vector is parallel to the quantum well plane, the situation is very similar to the bulk case because electrons can move freely along that direction. Therefore, the electro-optic effect resembles the Franz–Keldysh effect and has less practical interest. The most interesting case occurs when the electric field is applied along the direction perpendicular to the layers. In this case the effect is in some sense similar to the Stark effect in atoms, which produces a shift of the energy levels under the action of the field, and so it is called the *quantum confined Stark effect (QCSE)*.

Figure 8.11 shows the energy band diagrams for a quantum well before and after applying a constant electric field F perpendicular to the layers. The field introduces an electrostatic potential energy eFz for electrons, which adds to the crystal potential given by the profile of the conduction band minimum. Therefore, this profile tilts under the

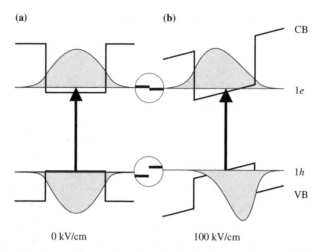

Figure 8.11. Band structure and wave functions for the fundamental states of electrons and heavy holes for a 100 Å thick GaAs/Al$_{0.35}$Ga$_{0.65}$As quantum well under electric fields of (a) zero field; (b) 100 kV/cm. Circular insets show Stark shifts of energy levels amplified by a factor 3.

influence of the field, as indicated in the figure and something equivalent occurs for the holes. The consequence of this perturbation is a deformation of the electronic states wave functions and a reduction of the confinement energy of carriers. As it corresponds to particles with a negative charge, the wave function of electrons deforms toward the direction opposite to that of the electric field. In contrast, the wave function of holes deforms toward the opposite direction. Due to the decrease of the confinement energy for electrons and holes the absorption edge of the quantum well shifts toward lower energy values, i.e. red-shifts.

Figure 8.12 shows the effect on the absorption spectrum of various electric fields perpendicular to the layers for a 9.4 nm thick GaAs quantum well [9]. In contrast to the bulk case, excitonic effects are important for field values up to 100 kV/cm. This is due to the fact that the well potential avoids exciton ionization by keeping the electron–hole separation smaller than the well thickness. The figure also shows the red-shift of the absorption edge and the exciton peak. Let us consider a rectangular quantum well with infinite barriers and a width L. For moderate electric fields the Stark shift ΔE can be easily obtained analytically from the Schrödinger equation by considering the applied potential as a second-order perturbation. For the fundamental level the shift is (see problem 7):

$$\Delta E_1 \approx -\frac{2.19 \times 10^{-3} e^2}{\hbar^2} m^* F^2 L^4 \tag{8.12}$$

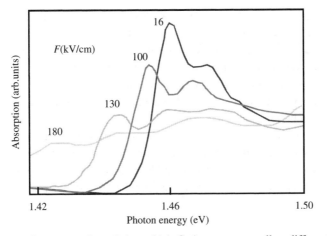

Figure 8.12. Absorption spectra for a 9.4 nm thick GaAs quantum well at different values of the electric field, applied normal to the layers. After [9].

where m^* is the effective mass in the direction perpendicular to the layers and e the electron charge. Note that the shift increases sharply with the well width and that it depends quadratically on the electric field, being independent on its sign (although this is valid only for symmetric wells). Moreover, due to the linear dependence with the effective mass, the shift is much larger for the heavy hole than for the electron or the light hole levels, and the electric-field-induced decrease of the quantum well bandgap is due mainly to the heavy-hole contribution. Figure 8.13 gives the Stark shift as a function of the applied

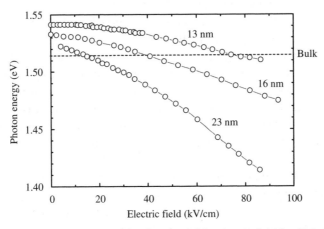

Figure 8.13. Excitonic band gap energy as a function of the electric field for GaAs/Al$_{0.35}$Ga$_{0.65}$As quantum wells of various thicknesses, obtained from photoluminescence emission experiments at 5 K. After [10].

field for GaAs quantum wells of different widths [10]. As an example, for a 100 Å-thick GaAs/Al$_{0.35}$Ga$_{0.65}$As quantum well under a field of $E = 100\,\text{kV/cm}$ Eq. (8.12) predicts shifts of -10 and -2 meV for the heavy hole and electron levels, respectively. A numerical calculation that takes into account the finite barrier height and the different effective masses in the barriers gives shifts of -15 and -6 meV, respectively. For photon energies slightly below the position of the latter at zero field, the well is transparent but shows a sharp increase on the light absorption when the field increases.

In summary, quantum wells can be used for the direct modulation of transmitted light beams. In this case, the electro-optic modulation is much more efficient than in devices based on bulk semiconductors or other conventional electro-optic materials like lithium niobate. Quantum well electro-optic modulators will be discussed in more detail in Section 10.8.

8.5. ELECTRO-OPTICAL EFFECTS IN SUPERLATTICES. STARK LADDERS AND BLOCH OSCILLATIONS

In coupled quantum wells, the electric field strongly affects the coupling between wells. Since the electrostatic energy of an electron in a uniform electric field depends linearly on the distance, the relative shift of levels originating in different wells is $\Delta E = eFd$, where d is the distance between the centres of the wells. The resonance between two states varies roughly inversely proportional to the energy separation of their levels and thus the field can easily tune the coupling among states coming from different wells.

When the coupling is small, an electronic state is localized mostly on one of the wells and the optical transitions can be classified as *intrawell transitions* among hole and electron states of the same well, and *interwell transitions*, among hole and electron states of different wells (Figure 8.14). For intrawell transitions the electro-optic effects are similar to those of single quantum wells. For interwell transitions the electric field induces a linear shift in energy $\Delta E = eFd$, which can be positive or negative depending on the direction of the field. If the latter points from the electron to the hole state, the transition experiences a red-shift. Interwell transitions usually decrease in intensity and finally disappear when the field grows. The linear shift in coupled quantum wells offers great advantages when a large tunability of the transitions is needed at moderate electric fields.

In superlattices the strongest excitonic peak corresponds to the transition from the top of the hole miniband to the bottom of the electron miniband (Figure 8.15(a)). When an electric field is applied normal to the layers, the coupling between the wells is reduced and the minibands split into a series of equally spaced discrete levels. The separation between two consecutive levels is $\Delta E = eFd$, where d is the superlattice period. The levels form the so called *Stark ladder*. Each level corresponds to an electronic state localized

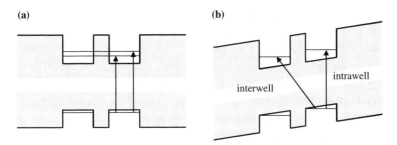

Figure 8.14. Symmetrical coupled double quantum well showing optical transitions from the right-well hole state. (a) At zero electric field the wells are coupled and electron states extend over both wells; (b) at finite electric fields the states are localized mostly in either one of the wells and transitions can be classified as intrawell or interwell.

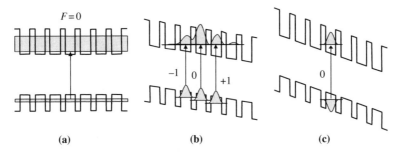

Figure 8.15. Electric field effects on a superlattice. (a) At zero electric field exciton transition occurs between miniband edges; (b) at moderate electric fields, holes are localized and the electron miniband splits into a Stark ladder, producing a number of transitions, intrawell and interwell; (c) at high electric fields electron states localize into single wells, and interwell transitions disappear. From (a) to (c) the absorption edge blue-shifts by roughly half the miniband width. For simplicity transitions are shown only to the central well electron state.

around a different well (Figure 8.15(b)). For fields of the order of $F \approx \Delta/ed$, where Δ and d are the miniband width and the superlattice period, respectively, the states of the Stark ladder become completely localized in the corresponding well (Figure 8.15(c)). This effect is difficult to observe in bulk crystals because the bandwidths are of the order of a few eV and extremely large electric fields are required. However, in superlattices the bandwidths are much smaller and the needed fields are easily attained. For example, in a typical 40/20 Å GaAs/Al$_{0.35}$Ga$_{0.65}$As superlattice the electron miniband has a width of about 65 meV and the electric field necessary for Stark localization is of the order of 115 kV/cm. The heavy-hole miniband has a width of only 6 meV and the heavy holes become localized for fields as small as 10 kV/cm.

The absorption spectrum of a superlattice (or related spectra, like photocurrent spectra, Figure 8.16) [11] reflects the field-induced localization. At zero electric field there is an exciton peak for the transition from the top of the hole miniband to the bottom of the electron miniband, corresponding to the transitions between delocalized states (Figure 8.15(a)). At intermediate fields (Figure 8.15(b)) the spectrum splits into a series of peaks corresponding to interwell transitions between the different states of the Stark ladder (Figure 8.17) [11]. At the fields where Stark localization occurs (Figure 8.15(c)) the interwell transitions disappear and only a peak corresponding to the intrawell transitions remains. Since the single well level is close to the centre of the miniband, the absorption edge of the superlattice experiences a blue-shift from zero to high fields. The blue-shift, which opposes the

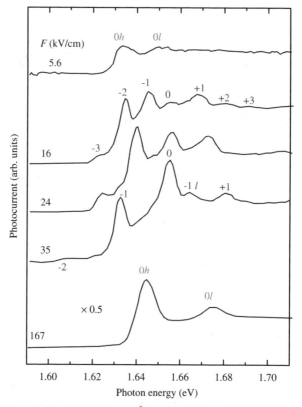

Figure 8.16. Photocurrent spectra for a 40–20 Å GaAs/Al$_{0.35}$Ga$_{0.65}$As superlattice measured at 5 K for various electric fields. (Spectra have been offset for clarity.) Interwell transitions are labelled according to the number of periods between the centres of the electron and the hole states. Note the blue-shift of the absorption edge from 0 to 167 kV/cm. After [11].

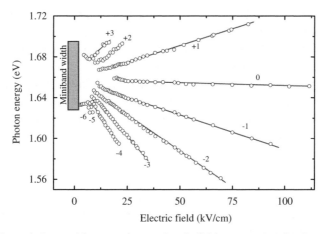

Figure 8.17. Heavy hole transition energies vs electric field measured under the same conditions as in Figure 8.16. Straight lines are a guide to the eye. After [11].

usual red-shift found in thick single quantum wells, is roughly equal to half the electron miniband width and has been the basis of some electro-optic devices.

In the previous section, we have limited ourselves to the static behaviour of electrons in the Stark ladder regime. However, the dynamical behaviour is also of great interest since it could provide the means to produce very fast electromagnetic emitters or *Bloch oscillators*. Terahertz emission was first observed in 1993, but it had already been predicted in the Esaki and Chang (1970) original paper on superlattices. Let us consider again the effect of a constant electric field on the electronic states of a superlattice with the help of Figure 8.18. At zero electric field quantum well states are fully coupled and the superlattice electron states have an infinite extension, their energy levels ranging continuously from the bottom to the top of the miniband (Figure 8.18(a)). Under a finite electric field the

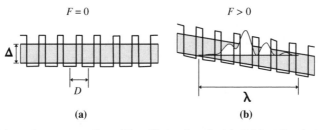

Figure 8.18. Schematic representation of the effects of an electric field on the electronic properties of a semiconductor superlattice. The field is applied perpendicular to the layers. (a) For a zero field; (b) for an applied electric field.

coupling is reduced and the miniband splits into a Stark ladder of levels (Figure 8.18(b)). The spatial extension of the corresponding states can be estimated by considering that the miniband is still a valid approximation. The edges of the miniband will tilt with the electric field due to the eFz term of the electrostatic potential energy. In a semiclassical picture the electron can only have kinetic energy values within the miniband range, thus defining a finite region of space where an electron at a certain energy level can move. The borders of this region are given by the values of z where the level energy intersects the edges of the miniband. The distance between these two points is:

$$\lambda = \frac{\Delta E}{eF} \tag{8.13}$$

In the absence of scattering, an electron wave packet will oscillate from one end to the other. As in the harmonic oscillator problem, the frequency is determined by the spacing between the levels, i.e.:

$$\omega_B = \frac{\Delta E}{\hbar} = \frac{eFd}{\hbar} \tag{8.14}$$

where d is the superlattice period. The angular frequency ω_B is called the *Bloch frequency* and can reach values in the THz range. Bloch oscillations have been observed in semiconductor superlattices by various optical techniques.

In summary, Bloch oscillations in superlattices can generate microwave radiation under the action of an electric field. The emission corresponds to transitions between the levels that form the Stark ladder. The emission occurs in the frequency range around 1 THz, which is not covered by conventional semiconductor microwave sources. However, efficient emission by devices based on superlattices still needs some more development before standard fabrication is implemented.

REFERENCES

[1] Fox, A.M. (1996) *Contemporary Physics*, **37**, 111.

[2] Cingolani, R., Tapfer, L. & Ploog, K. (1990) *Appl. Phys. Lett.*, **56**, 1233.

[3] Ekimov, A.I. & Efros, A.L. (1991) *Acta Physica Polonica A*, **79**, no. 1, 5.

[4] Empedocles, S.A., Norris, D.J. & Bawendi, M.G. (1996) *Phys. Rev. Lett.*, **77**, 3873.

[5] Goldstein, L., Glas, F., Marzin, J.Y., Charasse, M.N. & Le Roux, G. (1985) *Appl. Phys. Lett.*, **47**, 1099.

[6] Bimberg, D., Grundman, M. & Ledentsov, N.N. (1998) *MRS Bulletin*, **23**, no. 2, 31–34.

[7] Grundmann, M., Christen, J., Lendestov, N.N., Böhrer, J., Bimberg, D., Ruvimov, S.S., Werner, P., Richter, V., Gösele, V., Heydenreich, J., Ustinov, V.M.,

Egorov, A. Yu., Zhukov, A.Z., Kop'ev, P.S. & Alferov, Zh.I. (1995) *Phys. Rev. Lett.*, **74**, 4043.

[8] Canham, L.T. (1990) *Appl. Phys. Lett.*, **57**, 1046.

[9] Miller, D.A.B., Chemla, D.S., Damen, T.C., Gossard, A.C., Wiegmann, W., Wood, T.H. & Burrus, C.A. (1985) *Phys. Rev. B*, **32**, 1043.

[10] Viña, L., Mendez, E.E., Wang, W.I., Chang, L.L. & Esaki, L. (1987) *J. Phys. C: Solid State Phys.*, **20**, 2803.

[11] Agulló-Rueda, F., Mendez, E.E. & Hong, J.M. (1989) *Phys. Rev. B*, **40**, 1357.

FURTHER READING

Basu, P.K. (1997) *Theory of Optical Processes in Semiconductors* (Clarendon Press, Oxford).

Bimberg, D. Grundmann, M. & Ledentsov, N.N. (2001) *Quantum Dot Heterostructures* (Wiley, Chichester).

Bryant, G.W. & Solomon, G.S. (2005) *Optics of Quantum Dots and Wires* (Artech House, Boston).

Klingshirn, C. (2005) *Semiconductor Optics* (Springer, Berlin).

Schäfer, W. & Wegener, M. (2002) *Semiconductor Optics and Transport Phenomena* (Springer, Berlin).

Weisbuch, C. & Vinter, B. (1991) *Quantum Semiconductor Structures* (Academic Press, Boston).

Yu, P. & Cardona, M. (1996) *Fundamentals of Semiconductors* (Springer, Berlin).

PROBLEMS

1. **Optical transitions in a quantum well**. For the cases $n = 1$ and 2 in an AlGaAs/GaAs quantum well, calculate the transitions between the electrons and the holes, taking into consideration the selection rule of Section 8.2. Draw schematically the optical absorption coefficient as a function of photon energy. Assume the width of the well of $a = 10$ nm. *Hint*: observe that the absorption starts to increase strongly at the transition HH $\rightarrow$ e for $n = 1$, then decreases and increases again for the transition LH $\rightarrow$ e for $n = 1$. Almost simultaneously the increase due to the n_{2D} density of states starts. Continues to the $n = 2$ transition.

2. **Linewidth broadening in quantum wells**. One of the causes of linewidth broadening in quantum wells is due to the fluctuations in their width a. The influence of these fluctuations produces a band width increment Γ. (a) Show that, due to this effect, the broadening of the linewidth of the absorption or emission spectrum can be

expressed as

$$\Gamma = \frac{\hbar^2}{2} \frac{\pi^2}{a^3} \left(\frac{1}{m_e^*} + \frac{1}{m_h^*} \right) \Delta a$$

(b) Estimate the value Γ for a typical quantum well, assuming $\Delta a = a_0/2$, where a_0 is the lattice constant of the well material. (Observe that this situation is also encountered if one stops the production of the quantum well, by MBE techniques for instance, when the last monolayer is not completed.)

3. **Excitons in quantum wells**. From the results of Figure 4.16 discuss qualitatively the variation of the excitonic heavy hole and light hole energies with the well width. Discuss the differences in the values of binding energy for the HH and LH cases. Discuss also the differences for $x = 0.32$ and $x = 0.15$, where x indicates the proportion of Al atoms in the $Al_xGa_{1-x}As/GaAs$ structure.

4. **Quantum dots**. Consider a quantum dot made of GaAs of cubic shape with a size of 5 nm which has a quantum confinement energy of about 0.8 eV. Calculate the value of the Coulomb energy and show that is much smaller than the confinement energy.

5. **Si/SiGe superlattices**. Calculate the period of a Si/SiGe superlattice, so that the band minimum of Si, which occurs at a wave number of approximately $k \approx 0.8\pi/a_0$ (a_0 is the lattice constant), is brought to $k = 0$. *Hint*: according to the zone folding concept of Section 5.5.2, observe that the superlattices can behave optically as direct gap bulk semiconductors.

6. **Type II superlattices under electric fields**. In a type II superlattice (see Figure 5.7(b)), the electron and hole wave functions are separated by half the period, i.e. $d/2$. For a 80/30 Å AlAs/GaAs type II superlattice, calculate the energy shift produced by an electric field of value $F = 10^4$ Vcm^{-1}. Calculate also the Stark shift and observe that is smaller than the above result. (*Note*: it is known that when the thickness of GaAs is less than about 35 Å, the superlattices become of type II.)

7. **Quantum confined Stark effect**. Show Eq. (8.12) for the quantum confined Stark effect in a quantum well. *Hint*: take $H' = eFz$ as a perturbation Hamiltonian where F is the electric field applied along z (growth axis). Since the first-order shift gives zero as result, proceed to second order. Assume wave functions as if the quantum well had infinite walls. (*Note*: observe that this effect is similar to the quadratic Stark effect in the hydrogen atom.)

8. **Bloch oscillations**. Electron transport in a superlattice can be modelled by the electron motion in a weak periodic potential of period a. (a) Assuming that the electron scattering is negligible, calculate the frequency of Bloch oscillations for $a = 4$ nm and $F = 3 \times 10^6$ V/m. (b) Supposing a low concentration of electrons, explain how the polarization P corresponding to the electrons, i.e. $P = ex$, where x is the position and $v = (1/\hbar)(\partial E/\partial k)$ the group velocity, oscillates as a function of time.

Chapter 9

Electronic Devices Based on Nanostructures

Chapter 9

Electronic Devices Based on Nanostructures

9.1. INTRODUCTION

The high level of circuit integration in today's silicon technology could not have been achieved with III-V semiconductor compounds. However, from the point of view of operating speed, III-V devices show many advantages, mainly due to the high carrier mobility, μ, and lower effective mass of electrons in III-V compounds. As seen in Chapter 3 (Section 3.5.1), carrier mobility in GaAs is about one order of magnitude higher than that of silicon. In fact, the electron velocity in a semiconductor under the effect of an applied external electric field is probably the most representative parameter for the design of high-speed advanced electronic circuits. As we will see in this chapter, MODFETs, based on modulation-doped quantum heterostructures (Section 5.3.1), can work up to very high frequencies due to the large values of μ for parallel electron transport. The value of the cut-off frequency of these devices is larger than the corresponding values in Si-based MOSFETs and also GaAs MESFETs. It has to be pointed out that the high carrier mobility associated with these structures is a consequence of the quantification of electron states when a 2D system is formed, as well as the good quality of the AlGaAs–GaAs interface.

Figure 9.1 shows the maximum operation frequency (in GHz) of different MODFETs as a function of gate length (in microns) [1]. Due to their particular characteristics, these transistors are also called *high electron mobility transistor (HEMT)* (Section 9.2). For comparison purposes, Figure 9.1 also includes typical parameters of silicon MOSFETs as well as GaAs MESFETs. Frequency values are given for room temperature (300 K), although these frequencies are much higher for operation temperatures close to 0 K, as a consequence of the increase in mobility at low temperatures (Figures 6.2 and 6.3). At present, MODFET devices with gate lengths of about 100 nm and maximum operation frequencies at room temperature of several hundreds of Gigahertz (GHz) are available.

The use of quantum heterostructures is not only limited to field effect transistors, in which electron transport is parallel to the quantum well interfaces, but also to those transistors for which electron transport is perpendicular to the heterostructure interfaces. The operation of these transistors is based on the application of voltage differences to the emitter, base, and collector, similar to the case of bipolar junction transistors. In bipolar transistors, the maximum operating frequency is limited by the transit time of carriers through the base. As will be shown below (Section 9.3), *heterostructure bipolar transistors (HBTs)*, based on AlGaAs–GaAs junctions or on Si–Ge junctions, provide

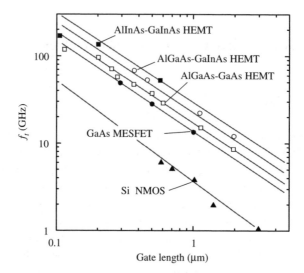

Figure 9.1. Maximum operation frequencies reached by MODFET (HEMT), MESFET, and MOSFETs as a function of gate length. After [1].

a notable improvement of the parameters of conventional silicon bipolar transistors, such as cut-off frequency, β factor, base resistivity, etc.

Another very interesting quantum effect to take into account for the development of advanced transistors is the so-called resonant tunnelling effect (Section 9.4). *Resonant tunnelling diodes (RTD)*, based on that effect, basically consist of a quantum well surrounded by two potential barriers thin enough to allow electron tunnelling. Due to the extremely low electron transit time through these semiconductor structures, electronic devices based on RTDs can operate at extremely high frequencies, in the range of 1 THz. By the addition of an RTD to a bipolar transistor or to a FET, it is possible to build *resonant tunnelling transistors (RTT)*, described in Section 9.6. In these transistors, the resonant tunnel structure injects very hot electrons (i.e. electrons of very high kinetic energy) into the transistor active region. Transistors based on this effect are called *hot electron transistors (HET)* and will be analysed in Section 9.5.

The reduction of the characteristic device size to the nanometric range leads to a notable reduction in the number of electrons contained in the electric signals transferred through electronic devices. This tendency has led to the development of the so-called *single electron transistor (SET)*. As will be seen in Section 9.7, the performance of SETs is based on the Coulomb blockade effect, which is manifested in zero-dimensional semiconductor structures, such as the so-called quantum dots (Section 4.6). The electronic current through a quantum dot in a SET, connected to the terminals by means of tunnel junctions, can be controlled electron by electron, by the application of a signal to an electrode that behaves as the gate of the transistor.

9.2. MODFETs

As previously seen in Section 5.3.1, a potential well for electrons is formed in AlGaAs–GaAs heterojunctions which, due to its reduced dimensions, results in quantified energy levels for the electron energy corresponding to the direction perpendicular to the interface. However, from the point of view of the motion parallel to the interface, electrons can be practically considered as free particles. It was remarked that the mobility of electrons can be, in this particular case, extremely high, since electrons which originated in the AlGaAs layer are transferred to the undoped GaAs layer where no impurity scattering takes place when they move under the action of an electric field parallel to the interface. These reasons motivated the fabrication from the beginning of the 1980s of very *high electron mobility transistors (HEMT)*. These transistors are also called *modulation-doped field effect transistors (MODFET)*, since they are based on modulation-doped heterojunctions, and their operation is controlled by an electric field that controls the motion of electrons along the channel. MODFETs, widely used for high-frequency applications, constitute a good example of devices whose performance is based on the quantum behaviour of electrons, since they are localized in a nanometric potential well which is smaller that the electron de Broglie wavelength (Section 1.3).

Field effect transistors based on heterojunctions show a layer structure that allows the creation of a 2D electron gas of high mobility. Figure 9.2(a) shows the cross-sectional schematic representation of a typical MODFET, with the source, gate, and drain electrodes. The representation of the energy bands, or more precisely, the conduction band in the perpendicular direction to the structure is shown in Figure 9.2(b). The most significant aspect related to these transistors is the potential quantum well for electrons formed between the n-doped AlGaAs semiconducting layer and the usually undoped GaAs layer. It has to be remembered from the AlGaAs–GaAs heterostructure (Section 5.3.1) that a potential quantum well is formed at the interface due to the larger gap of AlGaAs ($E_g \approx 2.0\,\mathrm{eV}$) in comparison to that of GaAs ($E_g = 1.41\,\mathrm{eV}$). The typical width of the quantum well, approximately triangular-shaped, is of about 8 nm, which is thin enough for the electron gas to behave as a 2D semiconductor. Figure 9.2(b) just shows one energy level. The role of the undoped AlGaAs spacer is to further separate the conducting electron channel from the n-type AlGaAs layer that generates the carriers, thus leading to higher electron mobilities due to reduced interaction with the ionized donors. The typical width of the spacer is about 50 Å.

It can be clearly appreciated from Figure 9.2 that the structure of a MODFET or HEMT is very similar to that of the MOSFET, already analysed in Section 5.2, in which the potential well for the electron channel is located at the Si–SiO$_2$ interface. In a similar way to the case of MOSFETs, in normal HEMT operation an electron current is created from source to drain when a voltage difference is applied between them. This current can be modulated by a voltage signal introduced through the gate lead. Likewise, the analytical

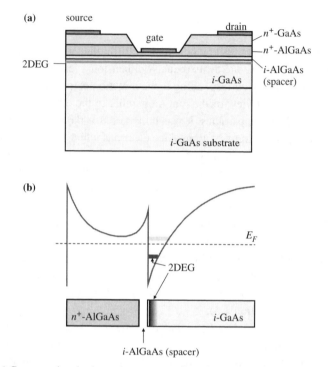

Figure 9.2. (a) Cross-sectional schematic representation of a MODFET (HEMT); (b) representation of the conduction band through a direction perpendicular to the structure.

expression that gives the dependence of the saturation current on the gate voltage, V_G, is in the form $I_{D,sat} \sim (V_G - V_T)^2$, where V_T accounts for the threshold voltage. As a consequence, the MODFET I–V characteristics are very similar to those of the MOSFETs. In relation to the performance of the MODFETs at high frequencies, the switching speed is improved as the electron transit time, t_r, is made shorter. Therefore the gate length, L, should be as short as possible ($L \approx 100$ nm). In contrast, a large width of the gate is usually preferred to increase the signal and the transistor's transconductance. In the case of MESFETs, in order to obtain large values of transconductance it is necessary to use highly doped samples (between 10^{18} and 10^{19} cm^{-3}), which limits the electron drift velocity due to impurity scattering by electrons. Hence, the MODFET structure presents an additional advantage since carrier transport takes place in the undoped layer (GaAs).

MODFETs dominate the low-noise device market, since they are capable of operating in the frequency range from microwaves up to about 100 GHz (Figure 9.1). The newly proposed AlGaAs–InGaAs–GaAs heterostructure makes electron confinement in the quantum well even more effective than in AlGaAs–GaAs heterojunctions and, in addition, electrons moving in the InGaAs layer show higher saturation drift velocities when compared to transport in GaAs. Transconductance reaches values up to 100 mS/mm, with cut-off frequencies of about 100 GHz and a noise level of only 2 dB. This behaviour is produced by the smaller gate-channel distance due to more abrupt barriers and reduced parasitic capacitance. For all these reasons, MODFETs are preferred in signal amplification in the microwave range, up to frequencies of 300 GHz, i.e. about six times higher than the maximum operation frequency of transistors fabricated with MOS technology, for a given lithographic resolution (Figure 9.1). MODFETs can also be fabricated from SiGe-based structures, although SiGe MODFETs have not moved into production like the HBTs (Section 9.3), due to their relatively high leakage currents.

9.3. HETEROJUNCTION BIPOLAR TRANSISTORS

A desirable property for junction bipolar transistors is to have a high value of the amplification factor β up to the largest frequencies possible. The maximum operating frequency of these devices depends on several design parameters, such as geometrical dimensions and doping levels of emitter, base, and collector regions. In order to obtain a high β, both the current gain through the base, α, and the injection efficiency factor of the emitter, γ, should be as close as possible to unity (it is assumed that the reader is acquainted with junction bipolar transistor terminology). This condition requires the emitter region doping level to be much higher than that of the base region. However, there is a reduction in the energy gap of semiconductors when the doping level is very high (Section 3.6). For instance, there is a gap reduction of about 14%, when the doping level reaches 10^{20} cm^{-3} which results in a notable reduction in carrier injection from the emitter region to the base region. For this reason, shortly after the invention of the homojunction bipolar transistor, Shockley suggested that the emitter could be fabricated by using a wide bandgap semiconductor. This would reduce the amount of injected carriers from the base region to the emitter region, thus improving the overall injection efficiency of the emitter. Bipolar transistors fabricated by using heterojuntions are called *heterostructure bipolar transistors (HBT)*, whose industrial production was started in the 1970s.

Figure 9.3(a) and (b) shows the differences between the band structures corresponding to npn homojunction and heterojunction transistors. Note that in the later case (Figure 9.3(b)), the band gap of the emitter is larger than that of the base region and consequently the barrier for the injection of electrons from the emitter to the base, eV_n, is lower than that corresponding to holes, eV_p, which results in a notable increment of the β factor. The barrier height difference has an enormous influence on carrier injection through the emitter–base junction, since injection processes show a quasi-exponential dependence on barrier height. In fact, the β factor is proportional to the ratio between the doping concentration of the emitter and base regions and to the coefficient $\exp(\Delta E_g / kT)$, where ΔE_g is the energy difference between the value of the wide bandgap of the emitter and that of the narrow bandgap of the base region. Thus at room temperature $kT \sim 0.026\,\mathrm{eV}$, a small gap difference ΔE_g substantially increases the β factor. All these factors mean that HBTs display an ample margin for fabricating heavily doped base transistors with a very low value of the base resistance and of the electron transit time through the base. Simultaneously, the doping of the emitter can be somewhat reduced, resulting in a smaller parasitic capacitance associated to the emitter–base junction. The simultaneous reduction of the base resistance and the capacitance of the emitter–base junction are essential for the correct performance of HBTs at high frequencies.

Another interesting feature of heterostructures is the possibility to fabricate a graded base HBT, with diminishing bandgap from the emitter to the collector region (Figure 9.4(a)). As a consequence, an internal electric field is created which accelerates electrons travelling through the base region, and therefore, allows HBTs to operate at even higher frequencies. Finally, if the collector region is also fabricated from a wide bandgap

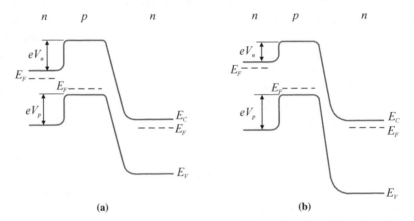

(a)   (b)

Figure 9.3. Band structure under polarization in the active region of: (a) homojunction bipolar transistor; (b) heterojunction bipolar transistor (HBT).

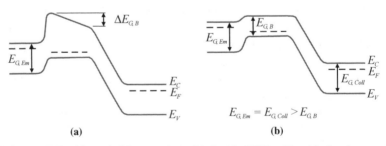

(a)   (b)

Figure 9.4. (a) HBT with graded base region; (b) double HBT, with wide bandgap emitter and collector semiconductors.

semiconductor (Figure 9.4(b)), the breakdown voltage of the base–collector junction can be notably increased. In addition, this structure, called double HBT or DHBT, allows the interchange between the emitter and collector terminals, which facilitates the design of integrated circuits.

Due to the good behaviour of AlGaAs–GaAs heterojuntions, and to high values of the mobility, HBTs are usually fabricated from III-V semiconductors. In a typical HBT, the base length can be of about 50 nm and is heavily doped, usually in the range of 10^{19} cm^{-3}. These transistors can be used up to frequencies of approximately 100 GHz, much higher than the corresponding values for silicon-based homojunction bipolar transistors. The use of InGaAs–InAlAs and InGaAs–InP based heterostructures allows even higher operating frequencies ($\sim$200 GHz) to be reached. An additional advantage of III-V semiconductor-based HBTs is the possibility of their integration in the same chip that includes electronic and optoelectronic devices. These *optoelectronic integrated circuits (OEIC)* usually include semiconductor lasers (Chapter 10), something which is not possible in the case of silicon-based technology.

There are also research projects focused on the development of HBTs based on silicon technology, which make use of different silicon compounds as wide bandgap materials. One of these compounds is silicon carbide (SiC), whose bandgap is 2.2 eV. Another material widely employed is hydrogenated amorphous silicon, whose bandgap is 1.6 eV. However, these materials show quite high emitter resistance, associated with the material itself or to the metallic contact. The most promising of all silicon compounds for the fabrication of HBTs are SiGe-based alloys, from which heterojunctions can be formed since the bandgap of silicon is 1.12 eV, while that of Ge is 0.66 eV. Si–SiGe heterostructure devices were developed much later (1998) than GaAs and other III-V materials due to the less mature SiGe growth technology. For the fabrication of Si and SiGe-based HBTs, the Si emitter region is usually followed by a SiGe base region with bandgap energy notably smaller than that of Si. This bandgap energy difference allows doping of the base region with relatively high concentrations, thus extending the allowed operation frequency to values comparable to those of III-V compounds. Commercial HBTs have cut-off frequencies over 100 GHz, while research devices have reached values close to 400 GHz. This high value of the cut-off frequency is partly due to compressive strains which change the energy band structure of the strained layers and, as a result, the effective mass of the carriers is reduced. Therefore, carrier mobilities are increased up to about 60%.

Finally, a gradual base region can be grown, by changing the value of x in the $Ge_x Si_{1-x}$ compound. The slope of the conduction energy band, as a consequence of the variation of the bandgap energy across the base, provides a quite high built-in electrical field ($\sim$10 kV/cm) which results in a reduced transit time for electrons as they travel through the base. These HBTs have a higher power dissipation than MOSFETs, but can be operated at higher frequencies and with lower noise. All these improvements make the SiGe-based HBTs very promising devices.

9.4. RESONANT TUNNEL EFFECT

It was shown in the previous sections that electrons in heterojunctions and in quantum wells can respond with very high mobility to applied electric fields parallel to the interfaces. In this section, the response to an electrical field perpendicular to the potential barriers at the interfaces will be considered. Under certain circumstances, electrons can tunnel through these potential barriers, constituting the so-called perpendicular transport (Section 6.3). Tunnelling currents through heterostructures can show zones of *negative differential resistance (NDR)*, which arise when the current level decreases for increasing voltage. The NDR effect was first observed by Esaki when studying p–n junction tunnel diodes in 1957 and, together with Tsu, proposed in the 1970s that this effect would be also observed in the current through quantum wells. However, it was not until the mid 1980s that the experimental growth deposition systems for heterostructures allowed the standard fabrication of quantum well devices based on NDR effects.

The operation of NDR quantum well electronic devices is based on the so-called *resonant tunnel effect (RTE)*, which takes place when the current travels through a structure formed by two thin barriers with a quantum well between them. The *I–V* characteristics of RTE devices are somewhat similar to that of Esaki's tunnel diode. Figure 9.5(a) shows the representation of the conduction band of a double heterojunction with a quantum well

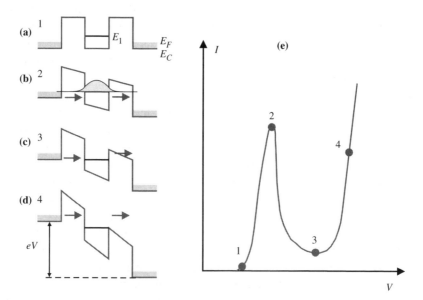

Figure 9.5. Schematic representation of the conduction band of a resonant tunnel diode: (a) with no voltage applied; (b), (c), and (d) for increasing applied voltages; (e) current–voltage characteristic.

between the junctions. The thickness of the quantum well is supposed to be small enough (5–10 nm) as to have only one allowed electron energy level E_1 (resonant level). The well region is made from lightly doped GaAs surrounded by higher gap AlGaAs. The outer layers are made from heavily doped n-type GaAs (n^+ GaAs) to facilitate the electrical contacts. The Fermi level of the n^+ GaAs is represented within the conduction band, since it can be considered a degenerated semiconductor (Section 3.6).

Let us suppose that an external voltage, V, is applied, starting from 0 V. It can be expected that some electrons tunnel from the n^+ GaAs conduction band through the potential barrier, thus resulting in increasing current for increasing voltage (region 1–2 in the *I–V* curve near 0 V). When the voltage increases, the electron energy in n^+ GaAs increases until the value $2E_1/e$ is reached, for which the energy of the electrons located in the neighbourhood of the Fermi level coincides with that of level E_1 of the electrons in the well (Figure 9.5(b)). In this case, resonance occurs and the coefficient of quantum transmission through the barriers rises very sharply. In effect, when the resonant condition is reached, the electron wave corresponding to the electrons in the well is coherently reflected between the two barriers (this is analogous to the optical effect produced in Fabry–Perot resonators). In this case, the electron wave incident from the left excites the resonant level of the electron in the well, thus increasing the transmission coefficient (and thus the current) through the potential barrier (region 2 in the *I–V* characteristic). In this condition, the effect is comparable to electrons impinging from the left being captured in the well and liberated through the second barrier. If the voltage is further increased (Figure 9.5(c)), the resonant energy level of the well is located below the cathode lead Fermi level and the current decreases (region 3), thus leading to the so-called negative differential resistance (NDR) region (region 2–3). Finally, for even higher applied voltages, Figure 9.5(d), the current again rises due to thermo-ionic emission over the barrier (region 4).

Commercial *resonant tunnelling diodes (RTDs)* used in microwave applications are based on this effect. A figure of merit used for RTDs is the peak-to-valley current ratio (*PVCR*), of their *I–V* characteristic, given by the ratio between the maximum current (point 2) and the minimum current in the valley (point 3). Although the normal values of the figure of merit are about five for AlGaAs–GaAs structures at room temperature, values up to 10 can be reached in devices fabricated from strained InAs layers, surrounded by AlAs barriers and operating at liquid nitrogen temperature.

If RTDs are simulated by a negative resistance in parallel with a diode capacitance C and a series resistance R_S, as is the case of normal diodes, it is relatively easy to demonstrate that the maximum operation frequency increases as C decreases. The resonant tunnel diode is fabricated from relatively low-doped semiconductors, which results in wide depletion regions between the barriers and the collector region, and accordingly, small equivalent capacity. For this reason, RTDs can operate at frequencies up to several terahertz (THz), much higher than those corresponding to Esaki's tunnel diodes which just reach about 100 GHz, with response time under 10^{-13} s. Small values of the negative differential resistance, i.e. an abrupt fall after the maximum of the I–V curve result in high cut-off frequencies of operation. In fact, RTDs are the only purely electronic devices that can operate up to frequencies close to 1 THz, the highest of any electron transit time device.

In a general sense, the power delivered from the RTDs to an external load is small and the output impedance is also relatively small. For this reason, it is sometimes hard to adapt them to the output of waveguides or antennas. The output signal is usually of low power (a few milliwatts) since the output voltage is usually lower than 0.3 V, due to the values of the barrier heights and energy levels in quantum wells. RTDs have been used to demonstrate circuits for numerous applications including static random access memories (SRAM), pulse generators, multivalued memory, multivalued and self-latching logic, analogue-to-digital converters, oscillator elements, shift registers, low-noise amplification, MOBILE logic, frequency multiplication, neural networks, and fuzzy logic. In particular, for logic applications, values of PVCR of 3 or higher and a high value of the peak current density, J_p, are required. In the case of memory applications, the ideal PVCR is 3 and values of J_p of a few Acm^{-2} are more appropriate. High frequency oscillators always require high J_p with PVCR over 2. Table 9.1 shows a comparison of the device performance of different materials systems.

Table 9.1. Comparison of RTDs from different materials systems. J_p is the peak current density, *PVCR* the peak-to-valley current ratio, $\Delta I \Delta V$ the maximum available power (assuming 100% efficiency) in the NDR region; and R_D the negative resistance of the diode in the NDR region. Adapted from Paul, D.J. (2004) *Semicond. Sci. Technol.*, **19**, R75–R108.

Material	InGaAs	InAs	Si/SiGe	GaAs	Si Esaki
J_p (kAcm^{-2})	460	370	282	250	151
PVCR	4	3.2	2.4	1.8	2.0
$\Delta I \Delta V$	5.4	9.4	43.0	4.0	1.1
R_D (Ω)	1.5	14.0	12.5	31.8	79.5
Area (μm^2)	16	1	25	5	2.2

9.5. HOT ELECTRON TRANSISTORS

When electrons are accelerated in high electric fields, they can acquire energies much higher than those corresponding to thermal equilibrium. For a three-dimensional crystal it is possible to associate a given temperature T_e, to electrons in the conduction band, by the use of relationship $E_k = (3/2)kT_e$, for the average kinetic energy of electrons. Evidently, in the case of a 2D electron gas the numeric factor of the above expression will be one. When the semiconductor is in thermal equilibrium, the electron temperature T_e coincides with the temperature T of the crystal structure. However, under non-equilibrium conditions, for instance, when an external applied electric field accelerates electrons to very high velocities, the kinetic energy, and thus T_e, can reach values which are much higher than those corresponding to the temperature of the crystal. In this case, the electrons are far from thermodynamic equilibrium, and receive the name of *hot electrons*.

Heterojunctions between different gap semiconductors allow the generation of hot elec-trons, since the electrons will acquire a kinetic energy, given by the energy discontinuity in the conduction band ΔE_c, when travelling from a wide bandgap semiconductor to one with smaller bandgap. In the particular case of the AlGaAs–GaAs heterojunction, the value of ΔE_c, ranges from 0.2 to 0.3 eV, which is about 10 times higher than $kT = 0.026$ eV at room temperature, and corresponds to a carrier velocity higher than 10^8 cms^{-1}. In addition, when travelling across the junction the electron beam suffers a collimat-ing effect as a consequence of the acceleration suffered by the electrons in the direction of the electric field, that is, perpendicular to the interface. This effect, called *electron injection by heterojunction*, makes the outgoing electrons concentrate inside a cone of about 10° of aperture.

One way of selecting the most energetic electrons in a given distribution consists of making them cross a potential barrier. Evidently, if the barrier is not very thin, only the most energetic electrons will have enough energy to overcome the barrier by a mechanism similar to the thermoionic effect. A much more effective procedure to inject hot electrons is based on the formation of thin potential barriers in the conduction band of semiconductor structures, to allow electron tunnelling. In this case, the resulting electron beam is almost monochromatic.

The idea of developing *hot electron transistors (HETs)* came when designing faster transistors, as proposed by Mead as early as the 1960s. However, these devices were not developed in any efficient way until some decades later, when growth techniques based on molecular beam epitaxy (MBE) allowed the fabrication of AlGaAs–GaAs heterojunctions of sufficient quality. Figure 9.6(a) shows the typical structure of a hot electron transis-tor [2], consisting of a n$^+$ GaAs emitter, a very thin ($\sim$50 Å) AlGaAs barrier, the GaAs base region ($\sim$1000 Å), another thick AlGaAs barrier of about 3000 Å, and the n$^+$ GaAs collector. When a positive voltage is applied to the collector, the injection of hot electrons coming from the emitter takes place by tunnelling through the thin AlGaAs barrier, since the base is positively biased with respect to the emitter (Figure 9.6(b)). It must be noted that the barrier's effective thickness can be modulated by varying the voltage difference between emitter and base, V_{BE}. The velocity of the injected electrons is, in this particular case, of about 5×10^8 cms^{-1}, much higher than in other type of transistors and, in addition, the electrons are collimated inside a cone of about 6°. The current gain through the base, α, can be made close to unity if both the scattering in the base region (which is usually very thin) and reflection in the collector barrier are reduced as much as possible. The base transit

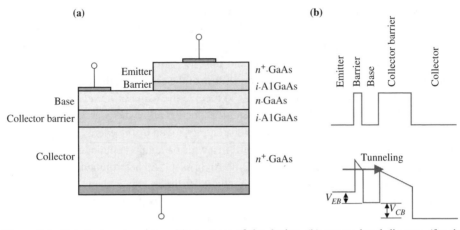

Figure 9.6. Hot electron transistor: (a) structure of the device; (b) energy band diagram (for the conduction band) under positive voltage applied to the collector. After [2].

time when the transistor is polarized can be of the order of tens of femtoseconds, but that associated to the crossing of the collector barrier is relatively higher. Nowadays, there are several efforts to reduce this time, although the collector barrier cannot be reduced as desired since leakage currents have to be prevented. Realistic estimations predict that the total time of transit should be considered on the order of 1 ps.

Similarly to the case of HEMTs, much effort is currently being developed to progressively reduce the dimensions of HET devices, so that the electron transit time is as short as possible. This evidently implies the reduction of the thickness of the space charge regions by increasing, for instance, the semiconductor doping level. However, this method has the drawback that the doping impurities can diffuse through the material and form complex chemical compounds that would result in a variation of the chemical composition of the materials. In order to overcome some of these problems, it was proposed to substitute the semiconductor at the base by a material that behaves as a metal, is non-contaminant, and does not show electromigration effects. The resulting device is called *metal base transistor (MBT)*. For the base region of MBTs, materials such as cobalt silicide ($CoSi_2$) can be used; this silicide shows a conductivity almost as high as that of metals and is chemically compatible with silicon technology. As is well known, the operation speed of bipolar transistors is limited by the low mobility of holes. In this sense, one of the most significant advantages of MBTs is that they are unipolar devices and can operate at higher frequencies.

The two most common MBTs structures are represented in Figure 9.7. Figure 9.7(a) shows the band structure of a device consisting of a metal-oxide-metal-oxide-metal heterostructure, under forward bias between the emitter–base and base–collector electrodes.

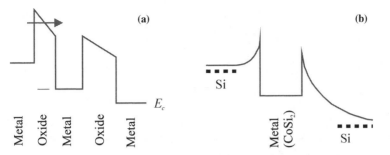

Figure 9.7. Energy band diagram of ballistic transistors: (a) metal-oxide-metal-oxide-metal heterostructure; (b) metallic base (cobalt silicide) heterostructure (Si–CoSi$_2$–Si).

In this case, electrons are injected by tunnelling through a thin barrier into the emitter junction. Figure 9.7(b) shows an even simpler MBT, formed by Si–CoSi$_2$–Si. The first Schottky junction is forward biased so that the electrons can overcome the emitter barrier by the thermoionic effect. CoSi$_2$ is chosen as material for the base region due to the good lattice matching properties with silicon that results in high quality interfaces and also due to its high electromigration resistivity. In these transistors, hot electrons behave as ballistic electrons when they reach the base region, that is, they practically do not suffer scattering since their mean free path is larger than the thickness of the base region.

9.6. RESONANT TUNNELLING TRANSISTOR

Diodes based on the resonant tunnel effect (RTE) discussed in Section 9.4 can be incorporated into standard bipolar transistors, field effect transistors or into hot electron transistors, thus creating devices with new properties called *resonant tunnelling transistors (RTT)*. Let us first consider a bipolar transistor in which a RTD is added to the emitter junction. Since the emitter to base polarization voltage, V_{EB}, controls the tunnelling resonant current, the collector current will show the typical RTD dependence (Figure 9.8(a)). Figure 9.8(b) shows the dependence of the collector current as a function of V_{CE}. Therefore, the output *I–V* characteristics present alternate regions of positive and negative transconductance that can be controlled by the voltage V_{EB}.

Figure 9.9 shows the band energy diagram of a transistor known as hot electron resonant tunnelling transistor biased in the active region [3]. Between the emitter and base regions of this transistor there exists a resonant tunnelling heterostructure, capable of injecting a large current when the electron resonant condition is reached. The position of the resonant level related to the emitter, is controlled by the voltage level applied to the base region, V_{BE}. This voltage can be increased until the resonant condition is reached.

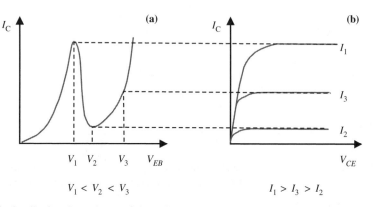

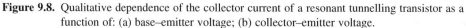

Figure 9.8. Qualitative dependence of the collector current of a resonant tunnelling transistor as a function of: (a) base–emitter voltage; (b) collector–emitter voltage.

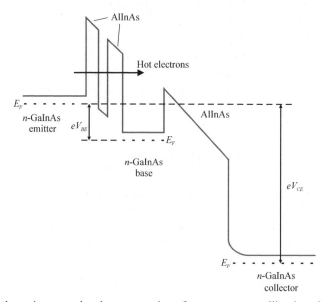

Figure 9.9. Schematic energy band representation of a resonant tunnelling hot electron transistor biased in the active region. After [3].

A maximum in the output current, I_C, is then produced. If V_{BE} is further increased, the current starts to diminish until a minimum value at V_2 is reached, similar to the description of the *I–V* characteristic of Figure 9.5 in Section 9.4. Therefore, the output charac-teristics of this transistor also show regions of negative differential resistance. Unlike simple HETs, the resonant tunnel structure injects electrons in a very narrow energy range

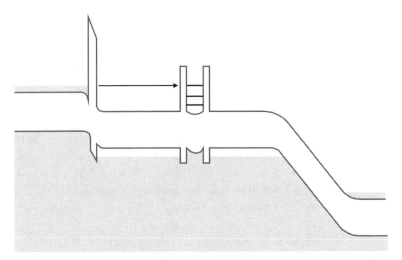

Figure 9.10. Schematic representation of a resonant tunnelling transistor (RTT) with a quantum well in the base region. After [4].

($\sim$1 meV, i.e. about ten times lower than that of HEMTs of Section 9.2). As pointed out above, resonant tunnelling diodes can also be incorporated in a different manner to bipolar transistors. Figure 9.10 shows a AlGaAs–GaAs bipolar transistor to which a RTD has been added to the base terminal [4], in addition to a tunnel injector (Section 9.5). Observe the quantum well between the two potential barriers in the RTD. The existence of several energy levels in the quantum well has been considered, which results in the apparition of various peaks in the collector current, rather than a single peak as in Figure 9.5.

There are several new applications of RTTs, mainly in the field of digital electronics. This is a consequence of the variation of I_C as a function of voltage curves as previously pointed out. In effect, let us suppose that several resonant tunnelling devices are connected in series with a voltage source V and a resistance R. The intersection of the load line with the characteristic curve of the tunnel device will give several stable operating points. If the number of stable points is two, we will have a binary logical circuit element. Evidently, if there are several energy levels available in the quantum well, we will have the same number of peaks or stable operating points. Digital amplifiers fabricated from these devices allow the implementation of logic gates with a smaller number of transistors than usually needed. For instance, a full adder circuit can be fabricated from just one resonant tunnel bipolar transistor and two standard ones, while the same conventional adding circuit needs about 40 transistors. As a consequence, much higher packing densities and computing speeds can be reached.

9.7. SINGLE ELECTRON TRANSISTOR

A very interesting aspect of future electronics is the capability of controlling with the highest possible accuracy the amount of charge in a tiny region, i.e. control the addition or the subtraction to the region of a single electron. The new field of single electron devices covers digital and analogical circuits, metrological standards, quantum information processing, etc. The concept of *single electron transistor (SET)* is based on the behaviour of 0D nanometric structures, such as quantum dots, in which electrons are distributed in discrete energy levels. One of the most interesting properties of these structures, associated to energy level quantification, is the so-called *Coulomb blockade effect*, which was already analysed in Section 6.4.3. When the tiny conducting material is extremely small (also called "island"), the electrostatic potential significantly increases even when only one electron is added to it.

In Section 6.4.3, we have seen that for the correct operation of SETs two conditions have to be met. First, the change in electric energy when an electron enters or leaves the quantum dot, i.e. the charging energy, has to be much larger than kT, which in terms of the capacitance is expressed as in Eq. (6.23): $C \ll e^2/kT$. Secondly, the resistance R_T of the tunnel junction must be large enough compared to the quantum resistance $R_Q = h/e^2$ ($\sim 25.8\,k\Omega$), Eq. (6.25), in order to avoid fluctuations in the number of electrons in the quantum dot as a consequence of the Heisenberg uncertainty principle.

At first, quantum dots were considered as two terminal devices. However, in order to fabricate transistors based on the Coulomb blockade effect, three terminals are needed. One of these terminals can be used as a gate to control the current flow through the quantum dot. Therefore, the SET basically consists of a quantum dot connected to the source and drain electrodes through tunnel junctions. The gate electrode is coupled to the quantum dot by an insulating material, in such a way that the electrons cannot tunnel through the barrier. Since the source or drain current flow is controlled by the gate, the described three-terminal device operates as a transistor, although it cannot be used for signal amplification. As it can be appreciated, the terminology used for the electrodes is similar to that in MOSFETs, the quantum dot playing the role of the channel region in MOSFETs.

Figure 9.11(a) shows the schematic representation of a SET and Figure 9.11(b) its equivalent circuit as a three-terminal device. The quantum dot, with total electron charge *Ne*, is widely referred to as *Coulomb island* and is connected to the source and drain by two tunnel barriers. The number of electrons in the Coulomb island can be controlled by the external voltage, V_G, through the equivalent gate capacitance of the semiconductor structure chosen to implement the transistor. Contrary to source and drain potential barriers, there is no tunnelling current through this electrode.

The current–voltage characteristics of the SET can be determined by applying a continuously sweeping voltage, V_G, to the gate electrode. The applied voltage induces a charge CV_G in the opposite plate of the capacitor, which is compensated by the tunnelling of a

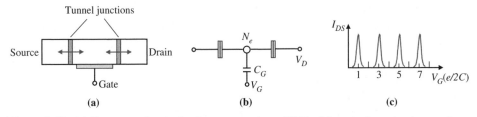

Figure 9.11. (a) Structure of a single electron transistor (SET); (b) equivalent circuit as a three-terminal device; (c) current as a function of the gate voltage.

single electron that enters the quantum dot. Thus, some kind of competence between the induced charge and the discrete one that tunnels through the barriers is established, that results in the so-called *Coulomb oscillations*, associated with the current flow due to the discrete charges that tunnel through the barriers. These oscillations are recorded as a current variation between the source and the drain, I_{DS}, as a function of the gate voltage, as represented in Figure 9.11(c). Between two consecutive peaks, the number of electrons in the quantum dot is fixed and therefore no current flows. The periodicity of the voltage peaks, ΔV, in these curves is given by the one electron variation in the total number of electrons in the quantum dot, i.e. $\Delta V = e/C$. In fact, the capacitance of the quantum dot can be obtained by measuring the voltage difference between two consecutive peaks.

From the point of view of specific applications, logic circuits based on SETs can be implemented, in particular, as inverters substituting CMOS transistors. The inverter will be based in the schematic representation shown in Figure 9.12(a) [5], in which each tunnel

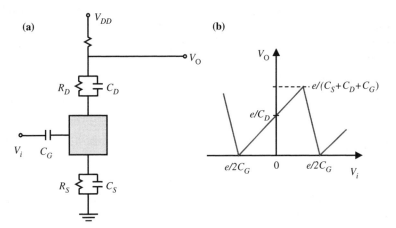

Figure 9.12. (a) Schematic representation of a SET as an inverter; (b) ideal transfer characteristic. After [5].

barrier is represented by a resistance in parallel with a capacitor. It is easy to observe that the output voltage, V_o, of this device as a function of the input voltage, V_i, shows a periodic variation, as shown in Figure 9.12(b). The periodicity of the output voltage is e/C_G and its amplitude is given by $e/(C_S + C_D + C_G)$. For this transistor, values for the voltage gain, $A_v = C_G/C_S$, close to 10 can be obtained, since $C_G \sim 1\,\text{fF}$ while $C_S = C_D \sim 0.1\,\text{fF}$. In practical devices, values of the gain of about 3 were demonstrated at low temperatures but not too close to 0 K. As T approaches 0 K, A_v should reach the theoretical amplification factor. The described inverters have been already used in the design of logical circuits and as unit memory cells formed by two inverters combined with MOSFETs.

Based on the conductance behaviour of carbon nanotubes, reminiscent of Coulomb blockade transport in metal and semiconductor wires and dots, carbon nanotube field effect transistors have been demonstrated for single- or few-electron charge-storage memories. The nanotube FETs show an extremely high mobility (about $10,000\,\text{cm}^2/\text{Vs}$), large geometrical capacitance, and large transconductance. In addition, it was recently demonstrated that single-walled carbon nanotubes FETs can be used in logic circuits, showing favourable device characteristics such as high gain (over 10), a large on–off ratio (over 10^5), and room temperature operation. Carbon nanotube transistor circuits exhibited a range of digital logic operations, apart from static random access memory cell, such as inverters, logic NOR, AC ring oscillators, etc.

Before ending the present chapter, a comparison of SETs and MOSFETs should be carried out. It is well known that until recently MOSFETs have been the basic devices of semiconductor chips, but we are already approaching their limiting feature size (Section 1.2). From this point of view, SETs would present the advantage of fabricating devices that, in principle, allow smaller sizes. They would also dissipate less power, since power consumption is proportional to the number of electrons in the input current flow to the device. One significant disadvantage of SETs is given by their high output impedance, due to the resistance associated to tunnel barriers. Another disadvantage is related to the size of the quantum dot, since for room temperature operation its capacitance should be as a small as possible.

REFERENCES

[1] Sze, S.M. (1991) *High Speed Devices* (Wiley, New York).

[2] Capasso, F. (1986) In *High-Speed Electronics*, Eds. Kallback, B. and Beneking, H. (Springer, Verlag, Berlín), **22**, 50–61.

[3] Yokoyama, N., Imamura, K., Ohnishi, H., Mori, T., Muto, S. & Shibatomi, A. (1988) *Solid State Electronics*, **31**, 577.

[4] Capasso, F. (1989) *IEEE Trans. Electron Devices*, **36**, 2065.

[5] Timp, G., Howard, R.E. & Mankiewich, P.M. (1999) In *Nanotechnology*, Ed. Timp, G. (Springer, New York).

FURTHER READING

Goser, K., Glösekötter, P. & Dienstuhl, J. (2004) *Nanoelectronics and Nanosystems* (Springer, Berlin).

Mitin, V.V., Kochelap, V.A. & Stroscio, M.A. (1999) *Quantum Heterostructures* (Cambridge University Press, Cambridge).

Modern Semiconductor Device Physics (1998) Ed. Sze, S.M. (Wiley, New York).

Nanotechnology (1999) Ed. Timp, G. (Springer, New York).

PROBLEMS

1. **MODFET.** (a) Compare the values of the carrier mobility, transit time under the gate, and maximum operating frequencies (Figure 9.1) of a MODFET and a MOSFET. (b) Explain why the cut-off frequency of a AlGaAs/GaAs MODFET is higher than that of a GaAs MESFET. (c) As shown in Figure 9.1, explain the better performance of an AlInAs/GaInAs transistor in comparison to the AlGaAs/GaAs transistor.

2. **Heterostructure bipolar transistor.** (a) Show that in a npn heterojunction bipolar transistor (HBT), the current gain β of the transistor is proportional to $\exp(\Delta E_g / kT)$, in the form,

    $$\beta \propto \frac{N_d}{N_a} e^{\Delta E_g / kT}$$

    where ΔE_g is the bandgap difference of emitter and base materials, N_d is the doping of the emitter, and the N_a doping of the base. (b) If typically $\Delta E_g \approx 0.2\,\text{eV}$ calculate the increase in β at room temperature. (c) Explain why even in a homojunction transistor ΔE_g is not zero. (*Hint*: the doping of the emitter is usually more than one order of magnitude higher than the doping of the base.) (d) Explain why in a HBT the emitter doping can be reduced to more realistic values and, in addition, the base resistance can be lowered.

3. **Hot electron transistor.** Suppose that electrons are injected from the emitter to the base in a AlGaAs/GaAs hot electron transistor similar to the one of Figure 9.6. (a) Show that the velocity v_B of the entering electrons at the base is given by $v_B \approx (2eV_b/m^*)^{1/2}$ where eV_b is the emitter barrier potential height. (b) Show also that

the electrons are collimated within a narrow velocity cone with a characteristic angle θ given by $\theta = (kT/eV_b)^{1/2}$. Calculate the value of θ at low temperatures, knowing that $eV_b \approx 0.3\,\text{eV}$.

4. **Resonant hot electron transistor**. Show how a resonant hot electron transistor (RHET) can be used as an "Exclusive-Nor" logic gate. *Hint*: consider the basic RHET of Figure 9.9. Put it to work in a common emitter configuration similar to that of bipolar transistors. The two binary inputs A and B are introduced in parallel to the base, while the result C of the logic operation is taken right at the collector.

5. **Real-space transfer of hot electrons**. Suppose the structure of Figure 6.5 formed by a quantum well material surrounded by two barrier materials, with mobilities μ_w and μ_b, respectively, and concentrations n_w and n_b respectively. (a) If $\mu_w \gg \mu_b$, show that the I–V characteristic presents negative differential resistance (NDR) and oscillations for high electric fields. *Hint*: consider spilling of electrons from the well to the surrounding barrier material when they acquire enough energy from the electric field to become hot. (b) Suppose we split the drain contact into two different contacts, one attached to the well and the other to the barrier semiconductor. Show how this three-terminal device can act as a transistor, and also that the current through the well would show a strong NDR and large PVR (peak-to-valley ratio).

6. **Single electron transistor**. Show that the transfer characteristic for the inverter configuration of the simple electron transistor (SET) of Figure 9.12(a) has the sawtooth form of Figure 9.12(b), where C_G represents the capacitances between the dot and the gate, and $C_\Sigma = C_S + C_D + C_G$ represents the sum of the capacitances between the source, drain, and gate. (Assume that $C_G = 8C_S$, $C_S = C_D$, $R_S = R_D \equiv R$). Explain why the period of oscillation of the gate voltage is q/C_G, and that the maximum voltage gain appears in the falling part of the curve and is given by $A_v = -|C_S/C_G| = 8$, while in the raising part takes the value $C_G/(C_D + C_G)$.

Chapter 10

Optoelectronic Devices Based on Nanostructures

Chapter 10

Optoelectronic Devices Based on Nanostructures

10.1. INTRODUCTION

This chapter is dedicated to optoelectronic and photonic devices based on nanostructures. Research into these devices has experienced a great upsurge during the last two decades due to the development of optical fibre communications. At present, there is a tendency to replace electronic functions by optical ones and to integrate electronic and optical devices in the same chip, creating the so-called optoelectronic integrated circuits (OEICs).

After the development of the semiconductor homojunction laser in the early 1960s, it soon became evident that double heterostructure lasers (Section 10.2) provided much higher carrier and optical confinement, which resulted in much lower threshold currents and better efficiency. More advanced semiconductor lasers, implemented in the late 1980s and based on quantum wells (Section 10.3), showed even lower threshold current densities ($\sim 50 \, \text{Acm}^2$), especially in the case of strained quantum wells (Section 10.5). Another type of quantum well laser considered in Section 10.4 is the vertical cavity surface emitting laser (VCSEL). Millions of these lasers, which emit from the top surface, can be integrated in the same chip, finding many applications in displays and optical signal processing.

Although not yet widely available commercially, we also consider in this chapter (Section 10.6) lasers based on systems with a dimensionality lower than 2D, like for instance quantum dots. These lasers, fabricated from 0D structures, show lower threshold currents and higher efficiency than quantum well lasers. However, with present lithographic techniques, there are still problems related to the difficulty of routinely producing quantum dots of small enough size and homogeneity. In this area, self-assembled growth techniques for the production of 0D and 1D structures are very promising. The last two sections of the chapter, Sections 10.7 and 10.8, deal with quantum well photodetectors and modulators, which are at present standard devices for long wavelength light detection and high-speed modulation of optical signals, respectively.

10.2. HETEROSTRUCTURE SEMICONDUCTOR LASERS

Before we treat laser structures based on low-dimensional quantum heterostructures, it is convenient to review some of the basic properties of semiconductor laser devices. We have seen in Section 3.7.5 that optical gain in semiconductors, as a consequence of stimulated emission, is obtained in $p^+ - n^+$ GaAs junctions of degenerate semiconductors

247

under forward bias. In this way, an active region with population inversion is created, since the Fermi quasi-levels of the degenerate p^+ and n^+ materials are located within the conduction and valence bands, respectively. Continuous laser operation is produced by the injection of carriers to the junction under forward bias. Lasers based on p–n junctions of the same semiconductor, as for example GaAs, have several drawbacks, partly due to the bad definition of the light emitting active region, with a size of about the diffusion length L_D, i.e. a few micrometre (Sections 1.3 and 3.5.3). In addition, the threshold current, i.e. the minimum current necessary for laser action, is quite large.

It soon became evident in the 1970s that *double heterostructure (DH) lasers*, which provide both carrier and optical confinement, could be much more efficient than homo-junction lasers and show threshold density currents ($\sim 1000\,\mathrm{Acm}^{-2}$) at least one order of magnitude lower. With these improved properties, DH lasers were very appropriate for applications in the emerging field of optical communications. Figures 10.1(a) and 10.1(b) show the basic structures of homojunction and DH lasers [1], respectively. The heterojunctions allow the formation of potential wells for electrons and holes, as shown in Figure 10.1(b), which increases the carrier concentration and, more importantly, the degree of inversion of the population of electrons and holes (Section 3.7.5). The active region of the DH lasers is still of the order of $0.1\,\mu$, a value not small enough to show quantized levels in the potential wells. Quantum well lasers will be treated in the following section.

One interesting additional aspect of DH lasers is due to the larger value ($\sim 5\%$) of the GaAs refractive index in comparison with the AlGaAs surrounding material. This difference is enough to provide an excellent optical confinement. The *optical confinement factor* Γ, which indicates what fraction of the photon density is located within the active laser region, is given by

$$\Gamma = \frac{\int_{\text{act.region}} |E(z)|^2\, dz}{\int_{-\infty}^{+\infty} |E(z)|^2 dz} \tag{10.1}$$

where $E(z)$ is the distribution of the electromagnetic wave amplitude in the perpendicular direction to the interfaces. In the DH lasers, the value of Γ can approach unity. Figure 10.1(b) shows the optical confinement effect in the active region for DH lasers.

In order to make the DH laser more efficient, the transverse *stripe geometry* configuration has been adopted almost universally (Figure 10.2). In this geometry, the transverse or horizontal dimension of the active region, and consequently the threshold current, is greatly reduced. Because of the shape of the active region, stripe geometry lasers are much easier to couple with fibres, waveguides, etc. The width of the active region in this geometry can be as small as $1\,\mu$m, so that the magnitude of the threshold currents is of the order of 10^{-2} A. In addition, optical confinement along the transverse direction can also be obtained by proper design of the index of refraction profile, in a similar way as it was done for the vertical direction. These lasers are called *index guided lasers* or buried

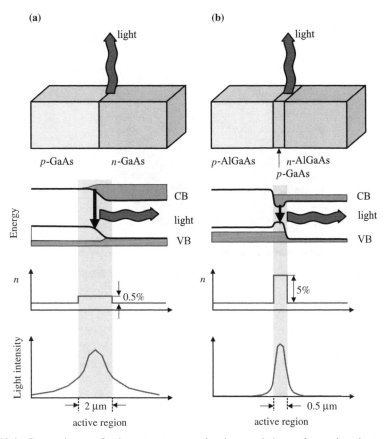

Figure 10.1. Comparison of the structure and characteristics of semiconductor lasers: (a) homojunction; (b) double heterostructure. From top to bottom: semiconductor regions forming the laser structure, band diagram which shows the potential wells for electrons and holes, profile of the refractive index, and optical confinement in the active region. After [1].

DH lasers (Figure 10.3). The width w of the optical Fabry–Perot cavity is made sufficiently small, so that only the lowest transverse modes of the optical radiation field will set up. Evidently, the transverse modes, for instance, TEM_{00}, will comprise several longitudinal modes, whose frequency separation depends on the cavity length.

Double heterostructures of the type n-AlGaAs–GaAs (active region)–p-AlGaAs–GaAs allow the creation of potential wells with excellent carrier and optical confinement as shown in Figure 10.1(b). Observe especially in this figure that potential wells for electrons and holes are created, a consequence of the values of the energy gaps of the semiconductors

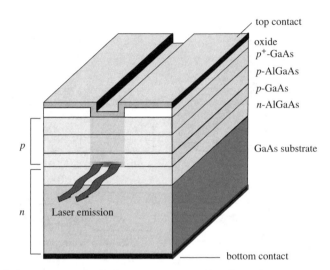

Figure 10.2. Stripe geometry double heterostructure semiconductor laser.

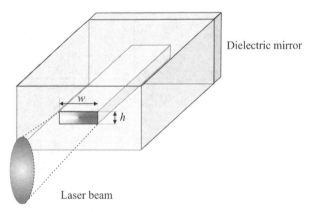

Figure 10.3. Geometry of the index guided semiconductor laser (it differs from the stripe geometry in the control of the optical confinement across the transversal direction).

forming the heterojunctions and the location of the Fermi levels. In spite of bandgap narrowing in degenerate semiconductors (Section 3.6), the wavelength of the light emitted by lasers based on AlGaAs–GaAs heterojunctions is still too short for transmission through optical fibres. Therefore, for optical communications, which take advantage of the 1.3 and 1.55 μm optical windows in fibres, one has to relay on quaternary InGaAsP on InP substrates (see Figure 4.9).

10.3. QUANTUM WELL SEMICONDUCTOR LASERS

As we have seen in the previous section, double heterostructure (DH) lasers show a high efficiency, low threshold current, and a high modulation bandwidth, which makes them very suitable for optical communications. In order to further improve their properties, and in particular to have a very narrow emission spectrum and wavelength tunability, the next natural step was the development of *quantum well (QW) lasers* in the late 1970s. In this section, we will see that the improved characteristics of quantum well lasers are mainly due to the properties of the 2D density of states function and characteristic of quantum wells. One drawback of the regular DH lasers is that both carrier confinement and optical waveguiding takes place in the same region. Figures 10.4(a) and 10.4(b) show two separate confinement structures frequently used. These structures are obtained for instance by grading the composition of $Al_xGa_{1-x}As$ compounds with values of x between zero and about 0.30, the value of the gap increasing from 1.41 eV to about 2.0 eV. In the separate confinement structure of Figure 10.4(a), a quantum well of width d of 5–10 nm is immersed in a wider cavity of width w which is optimized for enhancing the light waveguiding effect, thus having separate carrier and optical confinement structures. The waveguiding effect can be further improved by grading the refractive index, as shown in the lower part of Figure 10.4(b), in the so called *graded index separate confinement*

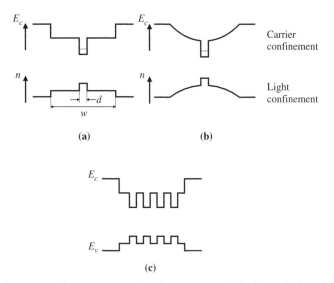

Figure 10.4. Separate confinement structures of quantum wells inside optical cavities: (a) profile of the conduction band and index of refraction; (b) GRINSCH structure; (c) multiple quantum well separated confinement heterostructure.

heterostructures (GRINSCH). Very often, in order to enlarge the emitted laser signal, a structure with *multiple quantum wells* is implemented (Figure 10.4(c)) instead of just one single quantum well.

Consider now a situation in which there is only one quantum well. In Section 3.7.5, we have derived the lasing condition $E_g < h\nu < E_{F_e} - E_{F_h}$ for bulk semiconductors, which means that the photons of the emission spectrum should have energies between the values corresponding to the gap, and that equal to the difference of the Fermi energy levels of the two degenerate semiconductors forming the heterojunction. The shape of the gain coefficient as a function of $h\nu$ was shown in Figure 3.18(b).

Let us now consider the *gain* for the 2D system and compare it with the case of the bulk, supposing lasing action between the $n = 1$ electron and hole levels. Evidently, the gain for the 2D system should start at $E_{g1} = E_g + E_{c_1} + E_{v_1}$ as can be appreciated in Figure 10.5,

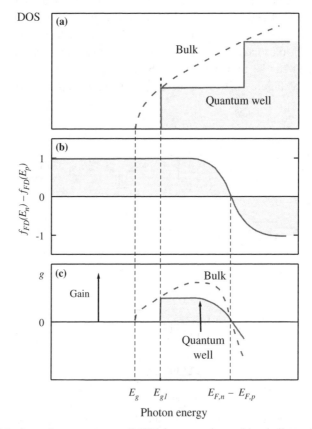

Figure 10.5. Gain factor in a quantum well (2D) in comparison with a bulk semiconductor (3D): (a) density of states functions; (b) occupancy probability factors; (c) gain factors.

while in the 3D case it starts at lower energies and increases much more slowly with energy. This is due to the fact that the electron density of states for the bulk increases only as the square root of energy above E_g, whereas the n_{2D} density of states function increases as a step function (Figure 10.5(a)). The first energy step of the distribution is considered, since the second step occurs at much higher energies. The important point to consider is the high slope (almost infinity) for the 2D system in comparison with the small slope for 3D. The decrease of the gain curves for the combined electron–hole system represented in Figure 10.5(c) arises from the Fermi factors (Figure 10.5(b)). Evidently, if we had considered the particular case of $T = 0\,\text{K}$, the curve of Figure 10.5(b) would go from $+1$ to -1 at exactly $E_{Fn} - E_{Fp}$ and the gain factor would have a constant value from $E_g + E_{c_1} + E_{v_1}$ to $E_{Fn} - E_{Fp}$. The concentration of the gain factor for the 2D system in a much narrower energy range than in the 3D case implies that the population inversion can be achieved with a much lower injected current in quantum well lasers in comparison to DH lasers.

Let us next indicate the steps for the calculation of the gain as a function of photon energy for various carrier concentrations, for the situation in which only one of the conduction and one of the valence subbands are occupied. For this, the quasi-Fermi levels are calculated from the concentration of injected electrons and holes as well as from the interband matrix elements. Next, the gain can be related to the current density J by assuming a value of τ for the recombination time, since $J = end\tau^{-1}$, where d is the thickness of the active region; alternatively, τ could be obtained from the rate of radiative recombination. The rate of non-radiative recombination as well as other loses, like leakage current, have to be subtracted. The results for the gain g_w in a quantum well are shown in Figure 10.6(a) [2]. For the case of MQW structures with n_w quantum wells, each with a gain g_w, the total gain $n_w g_w$ as a function of the total injection current density nJ is plotted in Figure 10.6(b) for $n_w = 1$, 2, 3, and 4. Observe that the gain starts to become positive in practically equal current steps which are proportional to n. The case of a single quantum well (SQW) can be compared to the MQW. If one is interested in small values of J_{th}, then SQW structure has advantages over the MQW. On the other hand, in order to obtain a high differential gain one should use a MQW structure.

QW lasers have proved to be very reliable and do not suffer degradation of the mirror facets as is the case for DH lasers, in which failure is induced by the large recombination rates in the active region. QW lasers also show a higher efficiency and smaller internal losses than DH lasers. Perhaps the main advantage for applications in high-speed optical communication is their high differential gain, defined as $G = dg/dJ$, previously mentioned, which for MQWs is of the order of $10\ \text{cm}^{-1}\text{mA}^{-1}$, about one order of magnitude larger than for the DH laser. For the high-speed operation of QW lasers, a proper design of the separate confinement well heterostructure is important. Towards this end, a built-in electric field can be produced (as in Section 9.3 for heterostructure transistors), which considerably decreases the effective times of capture and release of carriers inside the

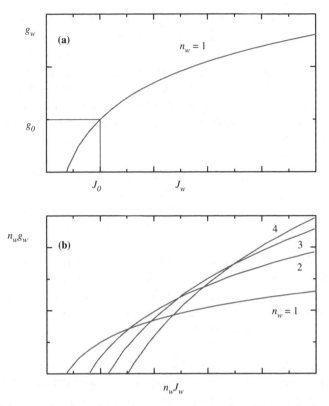

Figure 10.6. Gain as a function of injection current density for: (a) one single quantum well; (b) multiple quantum wells system with $n = 1$, 2, 3, and 4 single quantum wells. After [2].

optical and carrier confinement regions. These lasers can be modulated with injection currents up to frequencies of about 30 GHz.

10.4. VERTICAL CAVITY SURFACE EMITTING LASERS (VCSELs)

The most important characteristic of *vertical cavity surface emitting lasers (VCSELs)* is that light is emitted perpendicularly to the heterojunctions. There are several obvious advantages related to this geometry, including ease of testing at the wafer scale before packaging, the construction of large arrays of light sources (more than one million on a single chip), easy fibre coupling, and the possibility of using chip-to-chip optical interconnects.

The geometry of VCSELs, represented in Figure 10.7, consists of a vertical cavity along the direction of current flow (instead of perpendicular to it), in which the active regions surface dimensions are very small, so that light is extracted from the surface of the cavity rather than from the sides. Two very efficient reflectors are located at the top and bottom of the active layer. The reflectors are usually dielectric mirrors made of multiple quarter-wave thick layers of alternating high and low refractive indexes (for instance, GaAs/AlGaAs). The dielectric mirrors consist of *distributed Bragg reflectors (DBR)* which have a high selective reflectance at a wavelength λ given by (constructive interference):

$$n_1 d_1 + n_2 d_2 = \frac{\lambda}{2} \tag{10.2}$$

Since the reflectivity of these mirrors should be very high (close to 99%), as many as 30 layers are sometimes needed. The role of the high reflectance mirrors is to compensate for the low optical gain of the active region due to its short cavity length. Lateral confinement can be obtained with the mesa structure, as indicated in Figure 10.7, by etching through the upper DBR mirrors, sometimes even through the active region. Evidently, charge injection through DBR mirrors can only be possible if the materials are semiconductors. If insulators are used, such as titanium and silicon oxides, ring contacts can be employed to inject the charge directly into the active region. This region is usually quite thin, of the order of 100 nm, and consists of a few quantum wells, located at the maximum of the standing wave pattern produced in the middle of the two sets of mirrors. The set of quantum wells is surrounded by two spacers, one at each side, so that the vertical cavity

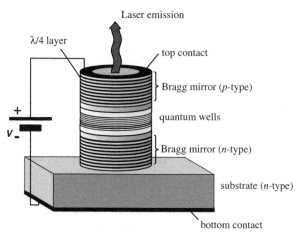

Figure 10.7. Schematic diagram of a vertical cavity surface emitting laser.

Figure 10.8. Array of VCSELs microlasers.

laser region has a length L such that $L \approx \lambda/n$. In this way, there is one full wave of constant amplitude in the active region and evanescent waves in the DBM mirrors, since these media are optically forbidden due to their high reflectivity.

One important aspect of VCSELs is that, due to the short cavity length in comparison to other "longitudinal" laser structures, the modes are widely separated. The frequency separation of the modes is of the order of 10^{13} Hz, comparable to the frequency at which the DBMs show their maximum reflectivity. Therefore, one can have single-mode oscillations. This is in contrast to edge-emitting lasers which have cavity lengths more than two orders of magnitude larger, and therefore, a small mode frequency spacing. Because of the small size of the resonant cavity dimensions, VCSELs are also called *microlasers*. Arrays of microlasers, as the one shown in Figure 10.8, with more than one million VCSELs in a microchip, provide very high optical power sources, which have important applications in optical communications and optical computation.

10.5. STRAINED QUANTUM WELL LASERS

Although the first quantum well lasers were constructed using lattice-matched heterostructures, especially from AlGaAs and InGaAsP compounds, it was shown some years later that strained layer lasers could show superior properties, specially from the point of view of lower threshold currents. Furthermore, strain introduces a new variable to extend wavelength tunability, in addition to controlling the width and barrier height of the quantum wells. The strained quantum well lasers were proposed in the late 1980s and developed in the early 1990s. Evidently, the strain depends on the amount of lattice mismatch as seen in Section 4.7. The lattice constant a of the ternary compound $In_{1-x}Ga_xAs$, frequently used in quantum well lasers, can be calculated by a linear interpolation from the GaAs and InAs lattice constants (Figure 4.9).

One of the most investigated strained quantum wells for lasers is the GaAs–InGaAs–GaAs. In this case, the inner InGaAs layer is under compressive strain. This has important consequences in the band structure (Figure 10.9) [3]. In addition to removing the degeneracy at $\vec{k} = 0$ between the HH and LH bands, as mentioned in Section 4.7, it changes considerably the values of the hole effective masses and increases the value of the energy gap. It is the large decrease in the parallel effective mass which causes a reduction in the threshold current. Compressive strains cause a reduction of a factor of about three of the in-plane (parallel) hole effective mass. This reduction makes the value of the hole effective mass similar to the electron effective mass, i.e. the curvatures of the $E = E(k)$ energy bands are almost the same for the conduction and valence bands.

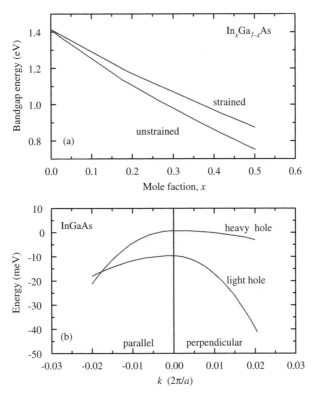

Figure 10.9. Strained InGaAs layers surrounded by GaAs: (a) In_xGa_{1-x} As bandgap as a function of composition; (b) heavy hole and light hole valence bands of InGaAs under compressive strain. After [3].

It can be shown [2] that in this case population inversion becomes more efficient, i.e. the gain coefficient increases as a function of the injected carrier concentration and the lasing condition occurs at lower threshold currents.

A variety of heterostructures are used for the construction of strained quantum well lasers with a wide tunability (0.9–1.55 μm): InGaAs–GaAs, InGaAs–InP, InGaAsP–InGaAs, InGaAlP–GaInP, etc. In addition to their low values of the threshold current, it has been verified that strained quantum well lasers show superior reliability, which seems to be due to the decrease of the propagation of defects in strained semiconductor layers.

The role of strained layers has been especially studied for quantum well lasers based in the materials system InGaAs–InP with a small bandgap energy appropriate for fibre optical communications. However, the values of J_{th} in experimental devices, even for the GRINSCH structure (Section 10.3), were larger than those predicted. This was solved by making use of strained active layers in this system, where values of $J_{th} \approx 200\,\mathrm{Acm}^{-2}$ have been obtained. In this case, the effectiveness of the strained layer has been attributed to the symmetrization of the conduction and valence bands. In order to find the correct value of the strain in the ternary alloy $In_{1-x}Ga_xAs$ on InP, the lattice constant $a(x)$ of the alloy can be expressed as a linear interpolation of those of GaAs and InAs, i.e. in Å, $a(x) = 5.6533x - 6.0584(1 - x)$. When InGaAs is grown over InP substrates with a lattice constant of 5.8688 Å, one can have either compressive or tensile strains depending on the value of x. For $x \approx 0.47$, the heterojunction is unstrained. At present, the physics and the applications of strained quantum well lasers are still under intensive investigation.

10.6. QUANTUM DOT LASERS

One of the most promising applications of semiconductor quantum dots and wires is for diode lasers. It was realized as early as 1976 that increasing the carrier confinement would provide multiple advantages for diode lasers over bulk materials. Among other advantages, the ideal quantum dot and quantum wire lasers would exhibit higher and narrower gain spectrum, low threshold currents, better stability with temperature, lower diffusion of carriers to the device surfaces, and a narrower emission line than double heterostructure or quantum well lasers. Real quantum dots and wires, however, have a finite size dispersion, defects, and limited carrier capture, which have slowed their progress. In the following, we will refer only to quantum dot lasers, since the experimental values obtained for quantum wire lasers are still far from the theoretical predictions. The growth technologies for quantum wire structures will have to improve, especially with respect to the quality of interfaces, uniformity of the wires, development of optical cavities with high confinement factor, etc.

In quantum dots, carriers are confined in the three directions in a very small region of space, producing quantum effects in the electronic properties. For optical applications, quantum dots must confine in the same region of space both electrons in the conduction band and holes in the valence band. The electronic joint density of states for semiconductor quantum dots shows sharp peaks corresponding to transitions between discrete energy levels of electrons and holes. Outside these levels the density of states vanishes. In many ways, the electronic structure of a quantum dot resembles that of a single atom. Lasers based on quantum dots could have properties similar to those of conventional ion gas lasers, with the advantage that the electronic structure of a quantum dot can be engineered by changing the base material, size, and shape.

Let us assume that the quantum dots are small enough so that the separation between the first two electron energy levels for both electrons and holes is much larger than the thermal energy kT (i.e. $E_{2e} - E_{1e} \gg kT$ and $E_{2hh} - E_{1hh} \gg kT$). Then for an undoped system, injected electrons and holes will occupy only the lowest level. Therefore, all injected electrons will contribute to the lasing transition from the E_{1e} to the E_{1hh} levels, reducing the threshold current with respect to other systems with lower confinement. The evolution of the threshold current density obtained along the years for various laser structures is shown in Figure 10.10 [4]. The lowest threshold currents have already been reached for quantum dot lasers. As long as the thermal energy is lower than the separation between the first and second levels, the emission band in an ideal quantum dot laser is very sharp and does not depend on temperature. Therefore, quantum dot lasers should have a better stability with temperature without the need for cooling.

The gain spectrum calculated for lasers based on different ideal quantum confinement structures is shown in Figure 10.11 [5]. Quantum dots should have the narrowest spectrum

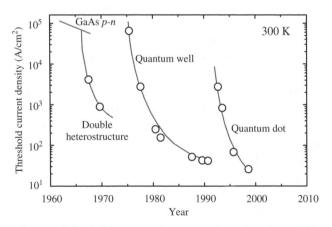

Figure 10.10. Evolution of threshold current density for lasers based on different confinement structures. After [4].

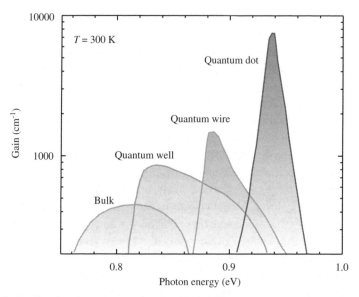

Figure 10.11. Predicted gain spectrum for the ideal bulk, quantum well, quantum wire, and quantum dot lasers. After [5].

and the highest gain. Also they should have a symmetrical spectrum, which would produce no jitter if the lasing wavelength is at the centre. For real quantum dots, however, there is a finite distribution of sizes that causes inhomogeneous broadening in the gain spectrum. This lowers the attainable gain, but has the advantage that the gain spectrum can be tailored by varying the size distribution. Tunable quantum dot lasers with a large tunability range are then possible. A more detailed treatment of the principles of quantum dot lasers can be found in the selected bibliography [6].

Despite the multiple advantages expected for quantum dot lasers, their development has been hindered by the difficulties found in the fabrication of arrays of quantum dots free of defects and with uniform sizes. Traditional methods for fabricating quantum dots include semiconductor precipitates in a glass matrix or etching away a previously grown epitaxial layer. None of these methods can produce large densities of dots, and the control of size and shape is difficult. Moreover they introduce large defects into the dots and create many surface states that lead to non-radiative recombination. Therefore, the appearance of quantum dot lasers had to wait until self-organized methods for the growth of quantum dots matured.

The most successful method to date has been the growth of *self-assembled quantum dots* at the interface of two lattice-mismatched materials. In this method a material such as InAs is grown by chemical vapour deposition, metalorganic vapour phase epitaxy or molecular beam epitaxy on a substrate with a larger lattice parameter and a larger bandgap such as GaAs. The first few monolayers grow in a planar mode with a large tensile strain and form the wetting layer. But beyond a critical thickness, it is more energetically favourable to form islands (the so-called Stranski–Krastanow regime) as shown in Figure 8.5 (Section 8.3.1). This creates a coherent array of pyramidal quantum dots on top of the wetting layer. Subsequently, a layer is overgrown epitaxially on top of the dots, creating an excellent heterostructure between two single-crystal materials: the dots and the surrounding matrix.

Figure 10.12 shows schematically an edge-emitting laser based on self-assembled quantum dots. The device consists of several layers forming a pin diode structure. The layers are, from bottom to top, the n-GaAs substrate, a n-AlGaAs layer, an intrinsic GaAs layer with the dots, a p-AlGaAs layer, and a p-GaAs cap layer. Metallic contacts on the substrate and the cap layer connect the device to an external circuit. Under a forward bias voltage, electrons and holes are injected into the middle intrinsic GaAs layer or active layer, where they fall into the quantum dots, which have a smaller bandgap, and recombine there. The emission wavelength corresponds to the interband transition of the InAs quantum dots. The GaAs layer, which is sandwiched between AlGaAs layers with a lower refractive index, confines the light and increases the interaction with the carriers. The InAs wetting layer contributes to an efficient diffusion of carriers into the dots. Its bandgap is smaller than that of GaAs, and therefore, collects carriers that reach the GaAs layer. Because the wetting layer is very thin, its bandgap is larger than that of the quantum dots, and carriers diffuse quickly into the dots. To increase the areal density of quantum dots, several wetting layers with pyramidal quantum dots are grown successively on top of each other with a layer of GaAs in between to form a stack of quantum dots.

The first Fabry–Perot laser based on self-organized quantum dots was developed in 1994 using InGaAs dots in a GaAs matrix [7]. At present quantum dot lasers emitting in the visible and infrared regions of the spectrum are already available in the market.

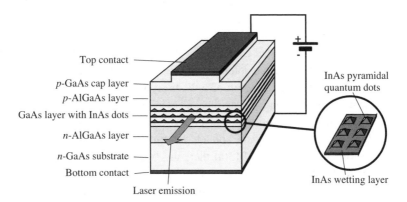

Top contact

p-GaAs cap layer
p-AlGaAs layer
GaAs layer with InAs dots
n-AlGaAs layer
n-GaAs substrate
Bottom contact

Laser emission

InAs pyramidal quantum dots

InAs wetting layer

Figure 10.12. Schematic illustration of a quantum dot laser based on self-assembled dots. The inset shows a detail of the wetting layer with the pyramidal quantum dots.

Although there is still room for improvement, these lasers already show some advantages over quantum well lasers. For example, their broader gain spectrum makes them useful for telecommunication amplifiers and tunable lasers. Also they present a better stability with operation temperature. Finally, GaAs quantum well lasers have an emission wavelength that is far from the fibre optic transmission windows. This spectral range has been covered traditionally with InP-based devices, but InP technology is not yet well developed. On the other hand, the growth of In-rich quantum wells on GaAs is difficult because of the large lattice mismatch. The advantage of InGaAs/GaAs quantum dot lasers is that they use the well established technology of GaAs substrates, but their emission wavelength falls within the fibre optic transmission window used for local area networks (1.3 μm).

10.7. QUANTUM WELL AND SUPERLATTICE PHOTODETECTORS

(a) Quantum well subband photodetectors

In principle, quantum wells can be used for the detection of light in any spectral region, as can be easily appreciated from their optical absorption properties (Section 8.2). However, it is in the IR region between 2 and 20 μm, that quantum well photodetectors are preferably used for example in applications of night vision and thermal imaging.

The problem with photodiodes based on band to band transitions across the semi-conductor gap E_g in p–n homojunctions is that they require materials with very low values of E_g, which makes it necessary to work at cryogenic temperatures. For instance, in the case of III-V compounds (see Figure 4.9) this leaves us with $InAs_{1-x}Sb_x$ with $x \approx 0.5$. Some II-VI compounds like HgCdTe can also be used in the IR, but these materials are quite soft, difficult to process, and have large dark currents. Another aspect which makes quantum wells very appropriate for use in IR detection is due to the large values of the dipole matrix elements corresponding to intersubband optical transitions (Section 8.2). In addition, wavelength tunability is easily implemented since the energies of the levels in a quantum well can be adjusted by the fabrication parameters, in particular its width. Especially important is the 8–12 μm interval in the IR, since in this wavelength range there is an atmospheric window, i.e. a region of low absorption which facilitates atmospheric optical communications, tracking of satellites, CO_2 laser optics (10.6 μm), etc.

Figure 10.13 shows the absorption transitions suitable for IR detection for a single quantum well under the action of an applied electric field, although practical devices are made with MQWs. In Figure 10.13(a) there are two energy levels in each well, the second level being located close to the top of the barriers. The separation between levels should be in the range 0.1–0.2 eV, which for III-V compounds implies a width of the wells of about 10 nm. We should remember (Section 8.2) that, due to the selection rules, the polarization of the incident radiation should be parallel to the confinement direction.

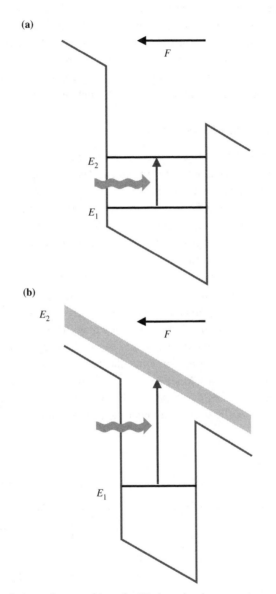

Figure 10.13. Optical absorption transitions for IR detection in a quantum well: (a) intersubband transitions; (b) transition from a bound state to the continuum narrow band outside the potential wells. (F is the applied electric field.)

Under light irradiation, this type of photodetectors generates a current by tunnelling of the carriers outside the wells. Sometimes, it is more effective to make use of absorption transitions between a single level in the well and the first continuum narrow band outside it (Figure 10.13(b)). In the case of the system AlGaAs–GaAs–AlGaAs, this energy is about 0.12 eV, and therefore, the spectral response is around 10 μm. The advantage of using this scheme is that the photodetector dark current is smaller than for the previous case in which the carriers had to leave the wells by tunnelling.

(b) Superlattice avalanche photodetectors

It is known that avalanche photodetectors (APD) based on semiconductors can present a high level of noise if precautions are not taken. The noise can be gradually reduced if the avalanche multiplication coefficient, α, is much larger for one of the carriers, for instance electrons, in comparison to the other carrier (hole) multiplication coefficient. In this sense, silicon is a very appropriate semiconductor for APDs, since the ratio α_e/α_h has a value of about 30. For a given semiconductor, the ratio α_e/α_h is practically fixed by the semiconductor band structure.

Quantum wells, on the other hand, allow a design control of α_e/α_h. For instance, a superlattice or MQW structure can be designed such that the conduction band discontinuities ΔE_c are much larger than the ΔE_v ones corresponding to the valence band. In this way, the electrons gain much more kinetic energy than the holes when they cross the band discontinuity. The same objective can be achieved by the design of a staircase profile superlattice (Figure 10.14(a)) for which the bandgap is graded in each well. In this case, the electrons have an extra kinetic energy ΔE_c when they enter the next quantum well. This extra energy makes the impact ionization phenomenon very efficient so that electron avalanches are easily generated under the action of an electric field F, as shown

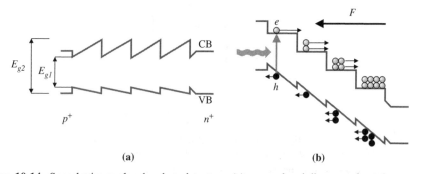

(a)   (b)

Figure 10.14. Superlattice avalanche photodetectors: (a) energy band diagram of a staircase superlattice; (b) formation of electron avalanche in the biased detector under light
irradiation.

in Figure 10.14(b). In contrast, holes only gain a small energy ΔE_v which is not high enough to produce impact ionization. Most superlattice APDs with very low noise figures are based on III-V compounds such as GaAs or InP. However, it should be mentioned that staircase superlattices are difficult to fabricate since their production requires strict control of the deposition parameters of quaternary III-V compounds.

10.8. QUANTUM WELL MODULATORS

Quantum wells can be conveniently used for the direct modulation of light, since they show much larger electro-optic effects than bulk semiconductors. Electro-optic effects are rather weak in bulk semiconductors and, for this reason conventional modulators make use of materials such as lithium niobate. In relation to excitonic absorption, we have already seen in Section 8.4 that, due to the quantum confined Stark effect (QCSE), large changes in the optical absorption spectrum of quantum wells could be induced by the application of electric fields. Because of the high barriers of the wells, excitons in these nanostructures do not field ionize as easily as in the bulk, and can therefore, sustain much higher electric fields ($\sim 10^5$ Vcm^{-1}). One important advantage of quantum well modulators is that they are compatible with microelectronic technology.

Electroabsorption modulators are based on the change of the optical absorption coefficient in a quantum well under effect of an electric field (Section 8.4). Figure 10.15 shows a mesa-etched modulator based on this effect. Evidently, to make the effect more

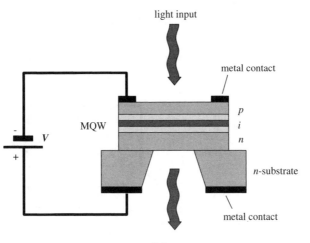

Figure 10.15. Mesa-etched electroabsorption modulator based on the quantum confined Stark effect.

significant, one uses a set of multiple quantum wells (MQW). The MQW structure consists generally of an array of several quantum wells (5 to 10 nm in thickness each) of the type AlGaAs–GaAs–AlGaAs. The structure is placed between the p^+ and n^+ sides of a reverse biased junction. Since the whole MQW structure has a thickness of about 0.5 μm, small reverse voltages can produce electric fields in the 10^4 to 10^5 Vcm^{-1} range. These fields induce changes in the excitonic absorption edge in the energy range 0.01–0.05 eV (see Figure 8.12).

Electroabsorption modulators, such as the one described, allow high speed modulation with a large contrast ratio of transmitted light through the device. The contrast ratio can be as high as 100 by working in the reflection mode instead of the transmission one. This is done by depositing a metal layer substrate and forcing light to make two passes. The modulation factor can also be improved by working at low temperatures. Electroabsorption modulators can operate up to frequencies of several tens of GHz and if high electric fields are applied, the maximum frequency can approach 100 GHz. This is because the maximum frequency of operation is limited by the mechanism of carrier extraction from the quantum wells. For low fields, the generated electron–hole pairs during absorption are unable to escape from the quantum well. However, if the fields are high enough, the electrons and holes can escape from the wells by tunnelling with a characteristic time of a few picoseconds.

Another use of quantum well modulators is also based on the QCSE (Section 8.4), but operating at a photon energy below the excitonic absorption edge. In this case, the electric field affects mostly the refractive index and consequently the phase of the light beam. The frequency of the photon energy should be close to the exciton absorption resonance to make the effect more pronounced, but not so close as to significantly absorb the light signal. For symmetric quantum wells, the dependence of the refractive index on the electric field is quadratic, like in the electro-optic or Kerr effect shown by centre-symmetric crystals in bulk semiconductors. However, in the case of quantum wells, the corresponding coefficient is about two orders of magnitude larger and consequently the

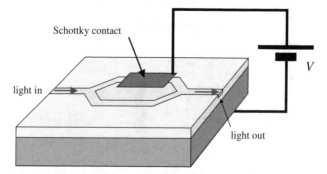

Figure 10.16. Schematic of a Mach–Zehnder interferometer.

distance which light has to travel through the material is much shorter. Therefore, devices with lengths shorter than about one hundred microns can be integrated in *optoelectronic integrated circuits (OEICs)*. One example is the *Mach–Zehnder interferometer* shown in Figure 10.16. In this device, the incoming signal from an optical waveguide is split in two beams of the same intensity which travel through different channels in the material of the same length before they recombine again. An electric field is applied to one of the branches causing differences in phase between the two beams and causing interference patterns at the meeting point.

REFERENCES

[1] Panish, M.B., Hayashi, I. & Sumski, S. (1970) *Appl. Phys. Lett.*, **16**, 326.

[2] Chuang, S.L. (1995) *Physics of Optoelectronic Devices* (Weley, New York).

[3] Zory, P.S.Jr, (1993) *Quantum Well Lasers* (Academic, Boston).

[4] Hatcher, M. (2000) *Opto-Laser Europe*, **74**, 33–34.

[5] Asada, M., Miyamoto, Y. & Suematsu, Y. (1986) *IEEE J. Quantum Electron*, **QE-22**, 1915.

[6] Bimberg, D., Grundmann, M. & Ledentsov, N.N. (2001) *Quantum Dot Hetero-structures* (Wiley, Chichester).

[7] Kirstaedter, N., Ledentsov, N.N., Grundmann, M., Bimberg, D., Ustinov, V.M., Ruvimov, S.S., Maximov, M.V., Kop'ev, P.S., Alferov, Zh.I., Richter, U., Werner, P., Gösele, U. & Heydenreich, J. (1994) *Electron Lett.*, **30**, 1416.

[8] Singh, J. (1995) *Semiconductor Optoelectronics* (McGraw-Hill, International Editions).

FURTHER READING

Bhattacharya, P. (1994) *Semiconductor Optoelectronic Devices* (Prentice Hall International Editions, Inc.).

Bimberg, D., Grundmann, M. & Ledentsov, N.N. (2001) *Quantum Dot Heterostructures* (Wiley, Chichester).

Chuang, S.L. (1995) *Physics of Optoelectronic Devices* (Wiley, New York).

Grundmann, M., Christen, J., Lendestov, N.N., Böhrer, J., Bimberg, D., Ruvimov, S.S., Werner, P., Richter, V., Gösele, V., Heydenreich, J., Ustinov, V.M., Egorov, A.Yu., Zhukov, A.Z., Kop'ev, P.S. & Alferov, Zh.I. (1995) *Phys. Rev. Lett.*, **74**, 4043.

Mitin, V.V., Kochelap, V.A. & Stroscio, M.A. (1999) *Quantum Heterostructures* (Cambridge University Press, Cambridge).

Yariv, A. (1997) *Optical Electronics in Modern Communications* (Oxford University Press, Oxford).

Zory, P.S.Jr., (1993) *Quantum Well Lasers* (Academic, Boston).

PROBLEMS

1. **Double heterostructure lasers**. Suppose an AlGaAs–GaAs–AlGaAs heterostructure laser which emits at 885 nm and that the optical length cavity is of 150 μm. (a) Find the order n of the principal optical mode (take the index of refraction of GaAs as 3.8). (b) Find the separation in wavelength and wave vector between modes. (c) For a temperature increase of 12°C, estimate the change in the emission wavelength and recalculate the order of the principal mode. (The index of refraction of GaAs changes about 1.5×10^{-4} per degree.)

2. **Optical emission in quantum wells**. For a typical AlGaAs–GaAs–AlGaAs quantum well laser of width 8 nm find the energy and wavelength of the emitted radiation. Compare the obtained value with that of a semiconductor GaAs laser. Solve the problem in the limit of infinite and finite quantum well heights.

3. **Distributed Bragg reflectors (DBR) and single mode lasers**. (A *DBR* structure consists in a periodic grating formed by a corrugated dielectric structure which is designed similarly to a reflection diffraction grating. *Distributed feedback lasers* (DFB) make use of DBR structures and are used in today's optical communications. They can be considered single frequency lasers and the spectral bandwidth is of the order of 0.1 nm.) (a) Find the Bragg wavelength λ_B of a DBR structure with a corrugation period of 240 nm and a refractive index of 3. *Hint*: first show that, if the DBR structure is considered as a grating, in first order, $\lambda_B = 2\ell_c n$, where ℓ_c is the corrugation period and n the refractive index. (b) Find the emission wavelength λ of the DFB laser for a length of the optical cavity of 300 μm. (*Hint*: in a DFB laser it can be shown that the only allowed modes are given by the expression $\lambda = \lambda_B + \lambda_B^2/2nL$ or $\lambda = \lambda_B - \lambda_B^2/2nL$, where L is the length of the optical cavity, and λ_B is the value obtained in (a).)

4. **Gain in a quantum well laser**. Show that the gain $\gamma(\hbar\omega_0)$ in a quantum well laser is proportional to the n_{2D} density of states function and to the inversion population factor $f_c(\hbar\omega_0) - f_v(\hbar\omega_0)$, where f_c and f_v are the occupancy Fermi–Dirac factors for the $n = 1$ energy states in the conduction and valence bands quantum wells, respectively. *Hint*: assume parabolic bands around $k = 0$ in k-space and that only the optical transition corresponding to $n = 1$ is significant. Follow the same procedure as in a bulk semiconductor and get

$$\gamma(\hbar\omega_0) = \frac{m_r^* \lambda_0^2}{4\pi \hbar a n^2 \tau} \left[f_c(\hbar\omega_0) - f_v(\hbar\omega_0) \right]$$

where m_r^* is the effective electron–hole reduced mass, a the well width, n the index of refraction, and τ the relaxation time.

5. **Quantum well modulators**. Consider a MQW electroabsorption modulator consisting of $Al_xGa_{1-x}As/GaAs$ quantum wells of width a equal to 8 nm and a total active layer thickness of $L = 640$ nm. The half-width of the excitonic peak is 25 meV. Calculate the modulation factor, i.e. the ratio of the intensities of the transmitted light with and without the action of an electric field $F = 10^5$ Vcm^{-1}. *Hint*: for the above quantum well, with $x = 0.3$ for the Al content, use the expression of the excitonic absorption given in Ref. [8]:

$$\alpha(\hbar\omega) \approx \frac{(2.9)(10^3)}{a\sigma} \exp\left(-\frac{E_{ex} - \hbar\omega}{(2)^{1/2}\sigma^2}\right)$$

where a should be written in Å, σ in eV, and α is given in cm^{-1}. Note that for no field $E_{ex} = \hbar\omega$, and when the field is applied the exponent ΔE_{ex} is given by Eq. (8.12) of Section 8.4.

6. **Electroabsorption modulators**. (Electroabsorption modulators, as the one shown in Figure 10.15 are said to be of the transverse transmission type. In them, an absorption coefficient, which is a function of the reverse applied voltage, i.e. $\alpha = \alpha(V)$, is defined as the average absorption coefficient characteristic of the MQW region. These modulators operate at photon energies $\hbar\omega_0$ close to the exciton peak; the variation of the peak intensity as a function of V can be obtained from curves similar to those represented in Figure 8.12 for various electric fields.) (a) Show that the contrast ratio R defined as the transmission for $V = 0$ and $V = V$, can be expressed in decibels as:

$$R(db) = 4.34 [\alpha(V) - \alpha(0)] W$$

where W is the width of MQW structure. Evidently, the contrast ratio increases with W. (b) Since W cannot be made very long because no signal would appear at the output, estimate what would be the optimum value of W so that the ratio P_{out}/P_{in} is not made too small.

Index